SpringerWienNewYork

Norbert Leitgeb

Safety of Electromedical Devices

Law – Risks – Opportunities

SpringerWienNewYork

Univ.-Prof. Dipl.-Ing. Dr. Norbert Leitgeb
Institute of Health Care Engineering
Graz University of Technology, Graz, Austria

SpringerWienNewYork is part of
Springer Science+Business Media
springer.at

Copy editing: le-tex publishing services GmbH, Leipzig, Germany
Typesetting and printing: C.H.Beck, Nördlingen, Germany

Printed on acid-free and chlorine-free bleached paper
SPIN: 12754021
Library of Congress Control Number: 2009943512

With 94 Figures

ISBN 978-3-211-99682-9 SpringerWienNewYork

Contents

Preface

Development in the field of medical technology has resulted in a manifold of medical devices enabling us to diagnose illnesses more reliably, treat them more efficiently and compensate for handicaps more effectively. However, these improvements are also associated with safety risks. Today, patients are in contact with an increasing number of medical devices longer and more intensively then before. Applied parts are put into contact with the body, probes may be introduced into the body via natural or surgical orifices, and even whole devices may be implanted for many years. The application of devices is no longer restricted to medical locations only. Home use by lay people is increasing and involves even critical devices such as for dialysis, nerve and muscle stimulation and ventilation.

In contrast to users' patients are in a special situation. Their life could depend on the performance of a device, they might be unconscious, may have impaired reactions, or have been made insensitive to pain by medication, and hence they may be exposed to hazards without their awareness and protection by their own reaction.

Therefore, medical devices must meet particularly stringent safety requirements. However, the question arises how safe is safe enough? The readiness to accept risks depends on a variety of accompanying circumstances. In fact, subjective risk perception varies among individuals and differs from country to country, and frequently only in rare cases it is in agreement with assessments of objective scientific analyses.

As a principle, total safety in terms of complete absence of any risk is not achievable. However, since safety is not available for free, the safety level accepted by a society is determined by a compromise between cost and benefit – or would you purchase a car regardless of its price, and only select the model that incorporates all achievable safety features?

Likewise, medical devices are not required to provide total safety. It is not even required that nothing severe shall happen. The objective of protection is solely that risk should be acceptable in relation to benefit – whatever this might mean. If, however, the situation is dramatic, if all alternatives have been tried and the last hope rests on a medical device that potentially could save a patient's life, even a high risk may be accepted in relation to the expected benefit. However, if more conservative methods were available, or the application would have only little relevance to health, risk assessment would be much stricter. As an example, a new method for blood pressure measurement, if associated with a lethal risk of thrombosis or cardiac infarct would not be acceptable in view of its limited benefit and existing alternatives with much less risk. However, who decides what risk can be imposed on a patient and what not?

Until recently the question was answered by standards that contained detailed safety requirements which were to be met by manufacturers. However, now the situation has considerably changed both in regard to legal restrictions as well as safety standards.

The new European medical devices directive 2007/47EC and the new edition of the international generic standard for electromedical devices EN IEC 60601-1 reflect this change. The safe but more restrictive way of defining particular safety requirements has been left behind, and now manufacturers have been guided onto the slippery parquet of individual responsibility. Now it is up to the manufacturer to define the safety level of a device under his sole responsibility, based on an implemented and maintained risk management process which is not restricted to just analysis and assessment of risks but comprises also further activities such as verification, validation, market surveillance and continuous evaluation and assessment of use experience. However, to accept this responsibility, manufacturers require particular knowledge to identify, imply and maintain the mandatory risk management process which must be maintained throughout the entire product life cycle.

However, in view of product liability, deficiencies in knowledge can become an existential risk. The reason is that manufacturers are liable also for consecutive damage caused by a product. In addition, the burden of proof has been reversed. Rather than be proven guilty, to escape from liability manufacturers must provide evidence for their innocence in terms of convincingly demonstrating that their *product* was not causally responsible for any damage.

The safety concept for medical technology involves also operators and users. It requires regular maintenance and recurrent safety testing by external and (if necessary) internal visual inspection and measuring and checking of safety-relevant parameters and performance. The change of the safety concept now challenges also testers and operators since they are no longer guided by particular requirements and standards but must try to understand the individual risk analysis of manufacturers when assessing the safety of a device.

This book aims at providing manufacturers, designers, safety technicians and operators with the general context and the essential framework of requirements for medical device safety. It describes which obstacles must be overcome, which pitfalls should be avoided, but also which opportunities exist in placing a medical device on the European market. It discusses which parameters influence individual risk perception, which safety objectives must be met and how the risk management process can be implemented including risk analysis, risk assessment, risk control and risk monitoring.

On the basis of a systematic description of recurrent safety testing, essential safety requirements are described. Step-by-step it is explained how external and internal visual inspection and safety measurements should be performed and by this approach basic knowledge is derived. The aim is to make the abstract wording of standards understandable and vivid.

However, it must be emphasized that this book does not aim at exhaustively discussing the numerous safety standards and legal requirements for medical devices. This is for three reasons:

(1) The concept is to provide easily understandable basic knowledge;
(2) exhaustive detailed discussion would have exceeded the practical limits of the book;
(3) standards are continuously changing, therefore too many details would soon become outdated.

Therefore, it is essential to be aware that this book aims at facilitating but not substituting working with standards. It is aimed at creating the required awareness for (safety) problems and giving a helpful overview to allow manufacturers and technicians identifying, estimating and assessing risks to derive responsible decisions for designing safe devices and performing reliable safety tests.

Graz, November 2009 Norbert Leitgeb

1 Medical devices

1.1 Background

Electromedical devices are different from universal electric appliances. Their application has consequences because they are intended to alleviate or heal a patient's disease or enable his/her further survival. Frequently, medical devices have to interact with or have to be introduced into the patient's body and – via applied parts – remain even for long periods in direct contact. In addition, patients may be unable to become aware of adverse situations or protect themselves by their reaction because of their condition, medications or illness. Therefore, they could be subjected to dangerous interactions even for long periods.

In contrast to other devices these aspects lead to specific risks, in particular due to

1. ***electric*** hazards, if the patient is a direct part of the electric circuit (e. g. ECG-monitor, nerve and muscle stimulator, defibrillator);
2. ***physical*** hazards due to insufficient solidity and/or stability (e. g. patient lifter) or noise (e. g. infant incubator), mechanical movement, pressure, overheating, fire, explosion or excessive radiation;
3. ***biologic*** hazards due to overdosage of highly effective drugs (e. g. infusion pump) or unintended release of adverse agents (e. g. allergenic, toxic or carcinogenic components) contained in material directly contacting intact skin, wounds or blood circulation;
4. ***hygienic*** hazards by transmitting pathogenic germs when touching contaminated parts (e. g. insufficiently sterilized endoscopes, catheters);
5. ***functional*** hazards through inaccuracies, malfunction and/or breakdown of life monitoring, supporting or sustaining devices (e. g. patient monitor, infusion pump, lung ventilator, pacemaker). Hazards may also occur by not providing claimed diagnostic or therapeutic effects (e. g. "miracle products" such as bioresonance devices) due to dangerously delayed application of methods of established efficiency.

For these reasons, it is generally accepted that medical devices need to be carefully designed, manufactured and maintained and have to meet tightened safety requirements. Today, this is demanded by European medical device directives /7/, /14/, /15/, /13/ and national medical device laws /53/.

Remark: European Union member states are obliged to transfer European framework laws ("directives") such as the directive of medical devices MDD 93/42/EC, active implantable medical devices IAMD 90/385/EC or in-vitro diagnostic devices IVD 98/79/EC into national law within an enforceable deadline.

By association agreements these regulations have been adopted also by countries outside the European Union such as Norway and Switzerland. In addition, based on mutual recognition agreements the European system of CE-marking and market approval of medical devices has been accepted already by many other industrialized countries such as Australia, New Zealand, Canada, Japan, Israel and partly by the USA.

These regulations require that medical devices are only allowed to be put on the market if they meet "essential requirements." Therefore, the obligation of a manufacturer to produce safe products is enforceable. Moreover, manufacturers are obliged to provide evidence of meeting these requirements. Depending on the risk potential of a medical device third-party verification by a European Notified Body may be necessary. This may be done by certifying production quality management (QM) and in addition by testing the device itself, its design, construction and the clinical evidence of the intended purpose.

1.2 What is a medical device?

Classification of a product as a medical device has consequences. These are legal, administrative and organizational. They include safety obligations in regard to risk management, safe design and production, market approval and additional requirements such as quality management, documentation and market surveillance – which therefore finally increase costs.

From the instructions for use:

The *Biovitalisator* can be used for haematomas and wounds, but may also accelerate fracture healing, helps against strain, rupture and cold hands, general inflammations, joint pain and arthrosis and supports leg drainage.
The Biovitalisator is a very good **wellness device**!

Therefore, a manufacturer has to decide whether his product is put on the market as a non-medical device (for healthy, conscious and reactive users) or as medical device (for the risk group of ill, unconscious, and non-reactive patients). As an example, the manufacturer can market a UV-irradiation lamp as a universal electrotechnical device (e. g. to harden bonds), as a cosmetic device (e. g. to tan the skin) or as a medical device (to treat skin diseases). He can declare a foot pedal device for fitness use or for medical diagnosis and therapy, or sell a magnetic mat as a wellness product or as a device for medical therapy. However, what a manufacturer is not allowed to do is to declare a product as a non-medical device and put it on the market under facilitated conditions and hence saving costs but at the same time market it by promising a cure for illness – as shown by the given example in the text box. Even the explicit declaration as a non-medical device does not change its medical device nature in the case of claimed medical indications.

Definition

Medical devices are manifold in regard to composition and complexity and may be instruments, apparatuses, software, material or other articles. In the European Medical

Devices Directive /7/, /14/ and in national medical device laws medical devices are defined as follows (citation /14/):

"Medical device means any instrument, apparatus, appliance, software, material or other article, whether used alone or in combination, together with any accessories, including the software intended by its manufacturer to be used specifically for diagnostic and/or therapeutic purpose and necessary for its proper application, intended by the manufacturer to be used for human beings."

Remark: Since the definition is restricted to human beings, devices intended solely for veterinary medicine do not fall under the regulations of the medical devices directive. However, restriction to human beings refers to market approval only. Technical requirements as laid down in European safety standards for medical electric devices /27/ refer to all patients, human or animal.

Remark: All medical devices, from plaster casts to heart – lung machines are regulated in the medical devices directive /7/, /14/ provided they do not fall into a special directive such as for in-vitro diagnostic devices /13/ or for implantable active medical devices /15/.

In the medical devices directive the purpose of use of a medical device is further specified. It includes application for

- ***diseases*** (diagnosis, prevention, monitoring, treatment or alleviation);
- ***handicap*** or ***injury*** (diagnosis, monitoring, treatment, alleviation, compensation);
- ***anatomy*** (investigation, replacement, modification);
- ***physiological processes*** (investigation, replacement, modification);
- ***control of conception*** (contraception or fertilization).

As an example, not considered a medical device is a hospital information system intended for patient data management. The reason is that it has no intended medical purpose while software for X-ray image processing or for ECG analysis does have a medical purpose and need to be classified as medical devices. Devices that are intended for research only (e. g. gene-chips or microarrays) are not considered medical devices provided the manufacturer does not declare them for medical use.

To distinguish medical devices from medicinal products their principal intended action must not be achieved by biochemical means such as pharmacologic, immunologic or metabolic effects. Therefore, substances with physical purpose such as bone cement, dental filling material, fibrin-based adhesives or contact lens cleaners are medical devices while substances administered to the body to treat or prevent diseases even with physical function, are considered medicinal products such as oxygen, infusion liquids, X-ray contrast agents or radiopharmaceuticals. However, if the main function of a device is based on physical effects and it is solely assisted by medicinal products, it remains a medical device which needs not be approved as a medicinal product – while the assisting pharmaceutical does need such an approval. Therefore, an empty syringe is a medical device (principle purpose: means to inject fluids into the body by applying mechanical pressure onto a piston), while a syringe marketed already filled with vaccine needs to be considered a medicinal product because its principal purpose is not injection as such but immunization of the patient while now the mechanical function of the syringe is only assisting in achieving the principle purpose of immunization. In contrast to this, an endoscope coated with a pharmaceutical (e.g. Heparin) to inhibit blood clotting remains a medical device since the medicinal product only assists its main purpose which remains dominated by the physical properties of the endoscope. Condoms (mechanical barriers) without or with assisting spermicide are medical devices while intrauterine devices (IUD) with integrated hormones are medicinal products. Medical gases for kryotherapy (e.g. CO_2, N or Ar), protective gases to prevent from explosions (e.g. N, NO, Ar) or gases intended to drive medical devices (e.g. compressed air, vacuum) are considered medical devices because they do not cause pharmaceutical effects (vacuum, which means low-pressure, is also named "medical gas").

Universal products without intended specific medical purpose will not become medical devices just because they are used within hospitals or medical premises. It remains the manufacturer and not the user who decides upon classification of a product and how to use it. Therefore, a hair cutter for preparing the head for brain surgery or a razor for removing hair from hairy skin to contact an RF (radio-frequency) surgery neutral electrode will not become medical devices just because of their use in medical surroundings.

> The manufacturer decides: He is the one who determines whether his product is a medical device or not. It is not the actual kind of application but solely the manufacturer that determines the kind of product by defining intended performance and use.

Accessories are considered medical devices if they are put on the market in their own right and if they are intended by the manufacturer to be used together with a medical device enabling its function (§ 2b MDD). However, accessories to accessories are no longer considered a medical device. For example, adhesive ECG electrodes for ECG monitors or infusion sets for infusion pumps are medical devices because the use of medical devices would not be achievable without them. Other examples are accessories to the medical product "medical gas" such as pressure gauges, decompression valves or gas connectors. Pacemaker electrodes are medical accessories while the screwdriver to fix them at the pacemaker is no medical product since it is an accessory to an accessory (electrode).

Spare parts are not medical devices and also not medical accessories provided they are not placed on the market in their own right and with a medical purpose. Therefore, indicator lamps, laser diodes, electronic components etc., are general spare parts even if they are used within medical devices. X-ray tubes for diagnostic and therapeutic devices may be spare parts but they may also be medical accessories if they are marketed for medical purposes.

Software is a medical product if it has a medical purpose in the meaning of the definition (§ 2a MDD). For example, software for treatment planning in radiotherapy, for biosignal analysis (e.g. ECG, EEG), for medical image processing or software controlling medical devices are medical devices. On the contrary, software for collecting and managing patient data or hospital information is not a medical device because the software is intended for administration and not for medical purposes.

Medical systems are considered medical devices even if they are placed on the market as a combination of several components to achieve their medical purpose (e.g. a suction pump system with pump, drain bottle, tube and cannula or an oxygen ventilator system with hose, respiratory mask and oxygen cylinder etc.). Use of universal products together with medical devices is not prohibited (e.g. ultrasound scanner with monitor, camera and video recorder). A power supply unit for a medical device can be an individual medical device, a component of a medical device or a (non-medical) device for universal use. In all these cases it could be used together with the medical device provided it conforms to its safety concept. An installed gas supply system consists of several universal products (e.g. gas pipes) and medical devices (valves, pressure gauges, connectors).

Medical procedure packs are medical products which are also placed on the market as a combination of several products which may not necessarily functionally interact but simply are intended to provide the user with all parts required for a medical procedure (e.g. surgical packages, first aid kits). Medical procedure packs may frequently be for single use.

As a matter of principle, it is the manufacturer who decides in which form he places a product on the market, whether he gives an own product name to an assembly of components to make it a medical system or a procedure pack or to separately market individually CE-marked products in their own right.

Demarcation aspects

Medical devices must be marketed by meeting special requirements and marketing rules. Therefore, demarcation from other types of products is important (Figure 1-1).

Not medical devices are (§ 5 und § 6 MDD /14/):

- ***personal protective devices***: They are governed by directive PPD 89/686/EG /16/. As an example, X-ray skirts are personal protective devices (for the staff) but when used as a means of gonad protection for the patient they are medical products. Surgical gloves protect (also) the surgeon. However, their principle intended use is prevention of disease for patients and surgeons. Therefore, their medical purpose dominates and they are considered medical devices but they must also meet (supplementary) all "essential requirements" of personal protective devices.
- ***cosmetics:*** They are governed by directive CD 76/768/EG /17/. As examples, UV-tanning devices or products for dental hygiene (e.g. dental brushes, oral irrigators, tooth paste or tooth brightener) are cosmetics and not medical devices.
- ***universal products*** are for general use without specific intended medical purpose such as general power supply units supplying medical devices. Therefore, operating system software or a computer does not become a medical device just because it is running medical software. Sun glasses are universal products while optical glasses are medical devices because they are intended to compensate for a disability. For example, a loupe or a digital camera with an attached monitor to magnify book pages and assist reading may still be considered a universal product because they are intended for general use with no specific dedication to the visually handicapped. The same applies to aids for opening cans or putting stockings on.

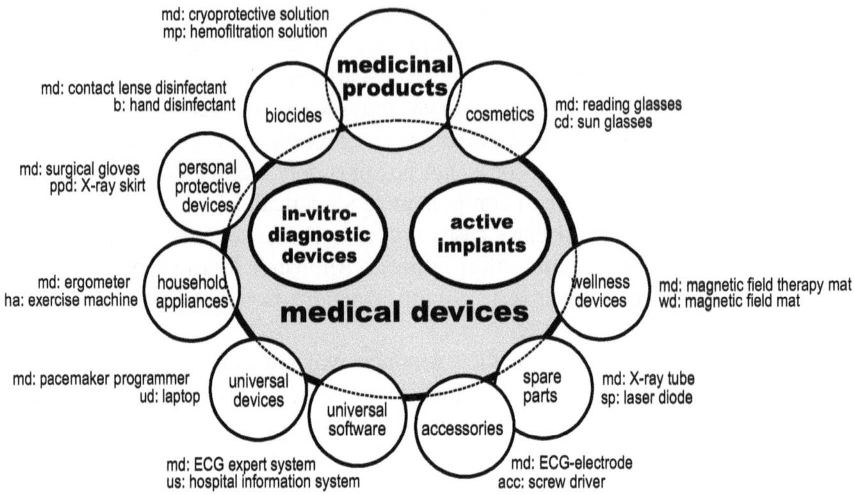

Figure 1-1: Demarcation of medical devices (md) to other types of products

- ***human*** blood, transplants, tissues or cells of human origin and devices containing human blood, plasma or blood cells are not medical products. However, medical products may contain or consist of derivatives of human blood or deactivated animal tissue.
- ***multipurpose disinfectants*** (together with other biocides) are governed by the biocides directive BD 98/8/EC and are not medical products although preventing diseases. However, disinfectants dedicated to medical devices such as endoscopes or contact lenses, are medical products.

Manufacturer

It has already been stressed that the manufacturer is the key. It is he alone who decides upon the intended purpose and the composition of his product, but he is also obligated to fulfil requirements imposed on him by the medical devices law. However, it is not always the case that the company putting a medical device on the market is also the company that designed and/or manufactured the medical device. It is no longer unusual that manufacturing is commissioned to a subcontractor and marketing is done by one or more vendors.

However, in regard to responsibility it is decisive only, who out of the chain of involved parties is declaring himself as the "manufacturer." The regulation is clear. The manufacturer with all legal obligations is solely that natural or legal person that declares itself as the manufacturer of the device. This is independent of who had really designed, manufactured or packaged the device (§ 2f MDD). Therefore, if a company imports a device, repackages it and solely marks it with its own brand, this is allowed. However, the company must be aware that it had become the new (sole) "manufacturer" who has to meet ***all*** requirements (including product liability) without restrictions (see Sects. 1.5, 1.6 and 1.7).

1.3 Which requirements must be met?

Medical devices are allowed to be put on the market and/or used on the patient only if they meet the "essential requirements" as defined in the European directives (and national medical device laws). They are defined in terms of general protection goals. Detailed requirements are formulated in supplementary European standards which have been declared "harmonized" with the directive by the European Commission. Compliance with essential requirements is assumed if such specific constructional and functional requirements of European standards are met.

Medical devices must be designed, manufactured, transported and stored such as to meet the following "essential requirements" (Annex I MDD /14/):

1. medical devices must exhibit an acceptable ***risk/benefit ratio*** when used under conditions and for the purpose intended by the manufacturer.
   To meet this, the manufacturer must implement a risk management process (according to EN ISO 14971 /21/). This should assure that risks of a device will sufficiently be managed by systematic risk analysis, risk assessment, risk evaluation, risk reduction and risk monitoring including risks caused by a user's foreseeable misuse, mistake or ignorance (see Chap. 2.2).
2. medical devices must be designed and constructed in conformity with generally acknowledged ***state of the art of science and technology*** and following the principles of ***integrated safety***.
   This means that even after safety testing and approval products are not allowed to be manufactured and marketed for unlimited time but require continuous adaptation to the actual state of the art. However, this does not require one to immediately implement any new finding. The generally acknowledged state of technology is stipulated by European standards. In addition, standards usually become binding only after perennial transit periods. However, no later than a set date a manufacturer must adapt construction and manufacturing to the new situation and needs to conform to the new requirements as defined or choose equivalent alternatives. This means that manufacturers have to actively review standards and other requirements, know them and (at least) meet their objectives.
   The requirement to follow the principles of "integrated safety" requires manufacturers to prefer efficient safety approaches (e.g. constructive) over less efficient (e.g., warnings). Therefore, protection against dangerous voltages needs to be assured by adequate insulation and not just by a warning not to touch live parts (see Chap. 2.3.3)
3. medical devices must ***achieve the medical performance*** intended by the manufacturer. Irrespective of the risk potential or conformity class of the product manufacturers must provide evidence for claimed performance of the device by clinically assessing medical knowledge and experience and/or performing clinical studies (see Chap. 3.2). Known problems are caused by miracle products of doubtable efficiency such as some devices used in alternative medicine.
4. medical devices must meet requirements 1–3 during their whole expected lifetime. This requires the manufacturer to decide upon the intended time of application. Limiting lifetime may be an element of his safety strategy. While expiry dates of

sterile products usually result from characteristics of the product and the durability of sterile packaging, the manufacturer may limit lifetime as a strategic decision allowing the use of ageing material such as rubber, to avoid maintenance or to manage risks from abrasion.

5. medical devices must fulfil their intended characteristics and performance also after storage and transport.
   This means that manufacturers have to consider also risks from distribution. It is the product's condition not just in the manufacturer's entrepots but at delivery to the client that is relevant (also for product liability claims). This requires adequate design of the packaging and definition of storage conditions and, if necessary limitation of storage time or lifetime.
6. medical devices must not have unintended side-effects constituting unacceptable risks.
   This needs to be checked by clinical assessment, based on existent knowledge and available experience (see Chap. 3.2). New applications may require checking by clinical studies.
7. medical devices must be accompanied by all information required for safe use during the intended lifetime.
   In general such information is pooled in instructions for use. They are an essential (and frequently underestimated) element to manage risk and liability (see Sect. 1.7.5 and Chap. 2.2). Attached information must consider the user's level of education in particular in case of lay use. The instructions for use must be written in an acceptable language with sufficient translation quality. In Germany and Austria, German is mandatory. Information may also be on the device or its package, but in any case must be contained also in the instructions for use. Required information comprises intended purpose, indications and contraindications, instructions for use, warnings, technical data, conditions of installation and handling, specifications for periodic inspection, testing and maintenance, lifetime (e.g. expiry date) and information on environment-conscious disposal. Manufacturers are not allowed to deprive information such as for periodic testing and maintenance by demanding such tasks be done only by personal authorized by them.
8. in addition, medical devices must meet several particular essential requirements concerning design, construction and performance (see Chaps. 3, 4, 5, 6, 7 and 8).

In addition to essential requirements of the medical devices directive, further essential requirements of other directives may be applicable if the medical device contains components which in their own right would be covered by these directive(s) such as

- the Machinery Directive MD 2006/42/EC /8/, if medical devices contain (even only parts of) machines such as drive systems, lifting accessories, mechanical safety components, load-handling components such as handles, lifting components such as chains, ropes and webbing or removable mechanical transmission devices. The additional essential requirements address construction, handling, control (including stopping), protection means, limited access, ergonomics, risks (e.g. mechanical, electric, thermal, noise, vibration and radiation), maintenance and additional information.

- the Directive on Personal Protective Equipment (PPE) 89/686/EEC /16/, if medical devices contain components or accessories intended to be worn or carried to protect from risks (e. g. X-ray skirts, laser protection glasses). Additional requirements concern design (ergonomics, safety level), additional risks (material, surface condition, hindrance), comfort and efficiency (adjustability, weight, rigidity) and information.

  Remark: If a device is intended for double purpose both as a personal protective device and as a medical device (e. g. gloves, masks, goggles), the entire conformity assessment procedure of either directives must be followed (including potential involvement of one or two notified bodies and adding both identification numbers to the CE mark).

- the Directive on Medicinal Products 2001/83/EEC /11/, if medical devices contain substances which, if used separately, are intended to diagnose, heal or prevent disease, or to re-establish biological functions. Medicinal components must be assessed by member states or the European Medicines Agency (EMA); their scientific opinion should consider quality and safety including risk/benefit ratio in regard to the specific intended use and potential degradations by the manufacturing process.

As a basic principle for all medical devices compliance with all essential requirements needs to be demonstrated by extensive documentation which, therefore, obviously must contain also quantitative test results. Depending on device-specific risks (or the conformity class (see Sect. 1.4.3), manufacturers may provide written evidence on their own or have to involve a European Notified Body.

1.4 How are medical devices placed on the market?

Initially, in other words before agreement on European-wide harmonization by medical device directive(s), market approval of medical devices and kind, contents and binding of safety standards differed from country to country. Therefore, depending on national requirements manufacturers had to meet time consuming and costly market-approval requirements including multiple third party device testing.

Remark: For instance, in Germany the medical devices safety standard was considered a technical rule which allowed deviations in justified cases. However, devices with increased inherent risk had to undergo a governmental design approval process which included mandatory type-testing by accredited test houses. In other countries such as France and Switzerland type testing was mandatory. In Austria, neither type testing nor governmental market approvals were mandatory; however, safety standards were made mandatory by listing them in ordinances to the electrotechnical law which made them non-negotiable requirements which needed to be met literally.

In a first step of harmonization contents of national standards were unified. A Europe-wide standardization agreement on the one hand offered the right to collaborate in the development of standards and to (weighted) voting, but on the other hand made it mandatory to adopt accepted European standards even in case of national denial. As a consequence, national standards now need to be withdrawn if they are in conflict with accepted European standards. Since the right of collaboration is associated with a national standstill, the development of individual national standards is only allowed in cases where a related inquiry on a New Work Item Proposal did not find transnational interests and therefore needed no involvement of CENELEC or CEN.

Remark: The obligation of withdrawal was the reason why national deviating medical gas colour codes existent in Germany, Austria, Switzerland and Hungary had to be adapted. After the end of a 10-year transit period the colour code of oxygen had to be changed from the former blue to the actual white. The colour blue is now assigned to nitrous oxide N_2O (laughing gas), which still constitutes some risk from confusing the meaning of colours.

As a consequence of this process, technical standards for medical devices are now harmonized throughout Europe and manufacturers can no longer refuse to eliminate defects by claiming them to be just a special request from a single country.

In spite of harmonized standard contents, because of their different legal binding trade barriers still persisted. For this reason, in 1986 the "New Approach" was initiated aiming at further removing these trade barriers by harmonizing legal bindings of standards and procedures for market approval throughout Europe. In a later further step, the "Global Approach," negotiations were started to harmonize requirements and approval worldwide. In the meantime, European and international standardizing bodies have coordinated their work. This resulted in elaborating common standards and paralleling voting procedures. For medical devices internationally harmonized standards are already available although without internationally harmonized binding. For instance, in the USA apart from accepted IEC standards such as the IEC 60601-series, further FDA (US Food and Drug Administration) and ANSI (American National Standards Institute) requirements are still existent, and local market approval is required and regulated by the "510k procedure" of the FDA /36/.

The European market approval procedure and CE-marking has been adopted by numerous European countries outside the European Union such as Croatia, Serbia and Turkey. In the meantime acceptance could also be achieved in important markets such as Australia, Canada and South-East Asia, and partial success has been reached in the USA (restricted to a list of devices).

Remark: FDA /36/ has established a classification of medical devices into three regulatory classes based on the level of control necessary to assure the safety and effectiveness of the device, namely
Class I: General Controls (with/without exemptions)
Class II: General Controls and Special Controls (with/without exemptions)
Class III: General Controls and Premarket Approval
Section 510(k) of the USA Food, Drug and Cosmetic Act requires manufacturers to register Class III devices and notify the FDA at least 90 days in advance of their intent to market such a medical device (Premarket Notification) for the first time or to reintroduce a device that has been significantly changed or modified.

1.4.1 Hierarchy of requirements

Elaboration of laws and standards has one thing in common: Both require time, sometimes even many years, during which technical possibilities continue to develop further. Therefore, there is a risk that regulations might be already outdated at the date of their issue. For example, until 1986 fuses in medical devices were not allowed to be soldered. This makes sense if easy and rapid change of the mains fuses should be made possible. However, in the course of technical development, many manufacturers began using fuses also to protect expensive internal electronic components such as microprocessors. Now fuses had no longer only the task of preventing hazardous overheating; some fuses were intended just to lower repair costs and consequently were soldered on the electronic board. In countries where the device standard was given a legal status, such as in Austria, this reasonable measure had to be objected because it violated the legal requirement (which had to be followed literally). Soldered fuses could be accepted only on the permission of the responsible ministry after passing a complex exemption procedure. Since such permission was restricted to a particular type of device only rather than accepting a principle solution, exemption procedures had to be repeated for other types of device. As a consequence, to the disadvantage of clients many manufacturers did not undergo this demanding procedure and just replaced such fuses by a short-circuiting clamp.

To allow manufacturers more flexibility to react to new developments more rapidly, in the meantime requirements for medical devices have been ranked in a hierarchical manner.

Directives

European Directives are in fact European laws rather than guidelines. They contain "essential requirements" which must be met absolutely and literally. Member states are obliged to transform European Directives into national laws within an enforceable deadline. Figure 1-2 summarizes in which way legal and technical requirements are elaborated and adopted.

European Standards

Since essential requirements just refer to generally formulated objectives, it is necessary to substantiate them in more detail. This is done by European standards. These describe an approved path to meet the essential requirements and are elaborated by the European standardizing bodies CENELEC (Comité Européen de Normalization Électrotechnique) and CEN (Comité Européen de Normalization) following the rules of the European standardization agreement. European standards are considered acknowledged rules of technology which have to be met as described or at least analogously. To allow reacting to technical progress it is permitted to deviate from them provided the standardized objectives remain met in another similarly effective way. However, in such a case manufacturers have to prove equality of their alternative solution.

Since the European standardizing bodies CENELEC and CEN cooperate with the international mirror organizations IEC (International Electrotechnical Committee) and

ISO (International Organization for Standardization) technical requirements are harmonized worldwide, although it is not mandatory to implement them nationally even in case of approval. This explains why, for example the USA agreed on these standards but kept individual regulations.

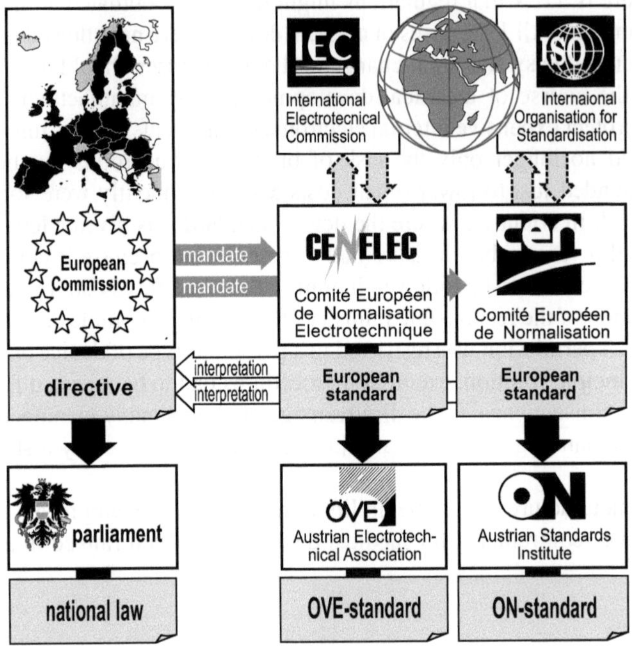

Figure 1-2: Development and transition of laws and standards

1.4.2 European market approval

Instead of different national procedures for market approval the European "New Approach" has implemented a system regulating market approval Europe-wide based on only one single procedure. Basically, a manufacturer has to perform the conformity assessment himself by checking conformity with the essential requirements. Independent third-party confirmation and certification of compliance with essential requirements has been assigned to European Notified Bodies. To signal conformity with European requirements the CE- (**C**onformité **E**uropéenne) mark (Figure 1-3) has been introduced. The initial objective was to further overcome trade barriers for CE-marked products.

Figure 1-3: European Conformity mark to overcome trade barriers

Conformity with essential requirements cannot only be assured by adequate design, construction and documentation. It also requires reliable, careful and reproducible manufacturing products. Therefore, in addition to the former requirements manufacturing needs to be governed by quality management systems. Manufacturers can choose one of four different options:

- ***outsourced final inspection*** *(Annex IV, MDD)*. This option disburdens manufacturers from implementing, maintaining and monitoring an own quality management system by outsourcing manufacturing final inspection to a European Notified Body. This option is especially useful for small- and medium-sized companies with discontinuous production since it causes calculable piece costs without requiring maintenance of costly QM activities with constant costs almost independent from sales. Outsourced final inspection can be performed by examining and testing each manufactured device or a random sample of a homogenous batch (which comprises a lot of devices usually with successive serial numbers, manufactured under constant conditions and in a continuous tperiod). After verification the notified body issues a written certificate of conformity for each individual device or the examined batch, respectively.
- own ***quality-assured final inspection*** *(Annex VI, MDD)*. This option requires implementing and maintaining an in-house quality management system restricted to final inspection which has been assessed, approved and certified, and is working under the surveillance of a European Notified Body which performs regular surveillance audits. The manufacturer must keep record, provide trained personal, calibrated testing equipment, working instructions and training schedules.
- ***quality-assured manufacturing*** *(Annex V, MDD)*. This option requires implementing and maintaining a certified and regularly audited quality management system comprising all manufacturing activities including final inspection.
- ***full quality management system*** *(Annex II, MDD)*. This option is the most complex one. It requires implementing and maintaining a certified and regularly audited quality management system comprising all product-related activities such as product development, design, construction, conformity assessment including testing, manufacturing, final inspection and market surveillance.

1.4.3 Medical devices differ

Medical devices are characterized by an enormous diversity. It reaches from non-critical reading glasses to high-risk life-sustaining heart – lung machines, from simple tongue depressors to complex magnetic resonance computed tomography imagers. In view of such diversity the effort of CE-marking and the rigor of third-party testing and verification by European Notified Bodies have been differentiated depending on product characteristics.

Therefore, depending on methodical risk, invasiveness, (uninterrupted) contact time and interaction with the patient medical devices are classified into four conformity classes I, IIa, IIb and III which can be roughly characterized as follows:

conformity class **I**	no or insignificant risk
conformity class **IIa**	small risk
conformity class **IIb**	elevated risk
conformity class **III**	high risk

Manufacturers are the primary decision-maker in regard of their medical device. They decide on the intended purpose of their device, its intended use and performance, and the way it is put on the market – and consequently they decide on the required effort for market approval.

Classification is performed by the manufacturer himself based on product characteristics, intended purpose and intended use as defined by him. Frequently manufacturers wish to achieve an exhaustive list of devices and their respective conformity classes. However, the conditions of use can be very different even for the same type of device. Therefore, such a list would be too unreliable. The reason is that even similar types of devices can easily fall into different conformity classes. As an example, suction devices could be conformity class I, IIa or IIb depending on whether they are intended for dental application (class I), surgical use (class IIa), or for bronchial aspiration (class IIb). Patient warning system may belong to class IIb if they are intended (also) for unconscious patients (e. g. during surgery), if not, they belong to class IIa. If ECG devices are just for recording, they are class IIa, if they are for monitoring (and alarming in critical situations) ECG devices are classified as class IIb.

Therefore, by extending or restricting the intended use (in the instructions of use) manufacturers have the possibility to decide on the conformity class of their device, and consequently, the efforts required for market approval.

Classification

Determination of the appropriate conformity class of a medical device is based on 18 complex classification rules laid down in Annex IX of the medical devices directive /14/. In addition, there are special decisions taken for certain borderline products (Figure 1-7). Depending on assessed device features classification rules can lead to different conformity classes. In such a case the highest identified class is applicable to the device. Classification is based on the following conditions:

1. for the ***intended use*** as defined by the manufacturer – and not for the additional technical possibilities or the actual application chosen by the user.
2. for ***worst case*** conditions of use as intended by the manufacturer – and not for intended use conditions of other comparable devices.
3. for ***normal*** conditions – and not for single-fault conditions (see Sect. 2.3.2). However, if a failure occurs too frequently, it is considered as a normal condition and no longer as a single-fault condition.
4. for ***all*** intended features of the product, even if leading to different classification pathways. The highest identified conformity class is assigned to the device.

Remark: If a manufacturer is uncertain how to classify his device, he can consult a notified body. In case of remaining uncertainties or different interpretation of classification rules the responsible "competent authority" might be asked for advice. If clarification cannot be reached at this level or in case of different national opinions the European Commission's Medical Expert Group may take the final decision.

5. for the ***marketed*** configuration. This means that the manufacturer can decide in which way the product is put on the market. He might wish to market his product assembled together with other products or accessories, or as a single product. As an example, he might market a suction device together with tube, cannula, and drain bottle as a medical system (entirely belonging to the highest identified conformity class) or to sell the various components separately as individual products in their own right, each with its own conformity class.
6. if it is marketed in its own right an accessory is classified according to its specific performance independent of the basic device.
7. software is classified as a self-contained product. If it influences the main purpose of a medical device (e.g. by controlling its function or modifying its results) it belongs to the same conformity class as the related device (Figure 1-4). As an example, a treatment planning software for X-ray therapy belongs to the same conformity class as the X-ray therapeutic device, in particular IIb; depending on its intended use ECG analyzing software belongs to class IIa or IIb, respectively; software for measuring the biparietal skull diameter within the ultrasound image of a foetus is classified IIa similar to the device.

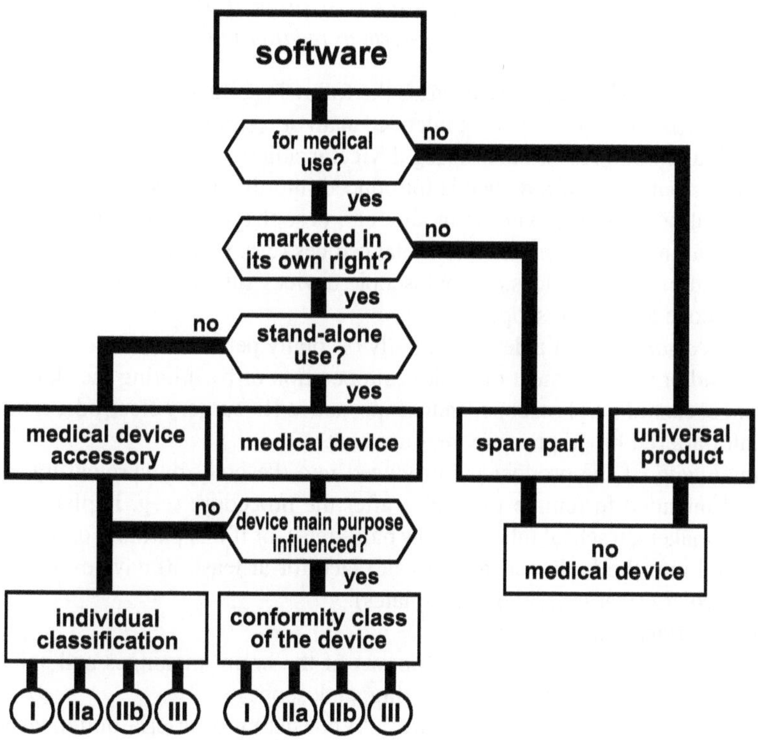

Figure 1-4: Paths to software conformity classes

To assess product properties the following main criteria are analyzed:

1. *methodical risk* to the patient. Methodical risks might be negligible (e. g. ergometer, infrared camera), small (e. g. ultrasound real-time scanner, ECG recorder), elevated (dialysis equipment, X-ray imager) and high (implanted cardiac pacemaker, implanted insulin pump).
2. *duration* of uninterrupted patient contact. It must be considered that the "assessment watch" is stopped each time contact is lost and restarted with each new contact. Interruptions are only neglected if they are needed to immediately replace a device with another identical one. For determining the conformity class contact duration is differentiated into
   - *transient* (continuous use for not more than 1 hour)
   - *short term* (continuous use for not more than 30 days)
   - *long term* (continuous use for more than 30 days)

   Because it is the uninterrupted contact that is relevant, for example surgical gloves are considered for transient use. The reason is that even during a several hours' surgery uninterrupted contact definitely remains far below 1 hour.

   Remark: This way of defining contact duration must not be confused with contact duration as defined for assessing biocompatibility (see Chap. 4). For this reason, it is the overall contact duration accumulated over the whole therapeutic process or even the whole professional life that counts. Consequently, the biocompatibility-relevant contact duration of surgical gloves in regard to the patient becomes short term and long-term in regard to the surgeon.

3. ***invasiveness***. In regard to invasiveness the following situations are differentiated:
   - *non-invasive:* if contact is not intended at all or restricted to intact skin only (e. g. manual blood pressure measuring device, irradiation devices).
   - *natural-invasive:* if a device is introduced into the body partially or totally via natural orifices (e. g. via the mouth, ear, oesophagus, trachea, rectum, urethra, vagina) including long-term artificial orifices (e. g. tracheal tubes). Examples of natural invasive devices are endoscopic devices such as the gastroscope, bronchoscope and proctoscope.
   - *surgical-invasive:* if a device partially or totally penetrates inside the body with the aid or in the context of a surgical operation or by injuring the skin (e. g. syringe, cannula, arthroscope, endoscope, active RF surgery electrode, catheter for intra-arterial blood pressure measurement).
   - *implanted:* if the product is introduced into the body by surgical intervention and intended to remain there also after the procedure (e. g. implanted cardiac pacemakers, tracheal tube, marrow nail, stent), or if it is partially or totally introduced into the body and remains in place for at least 30 days even if it is removed afterwards (e. g. fracture plates).
4. ***critical contact:*** such as with
   - *central nervous system* which consists of the brain, meninges and spinal cord (e. g. intracranial pressure monitoring, endoscope).
   - *central circulatory system* (Figure 1-5) consisting of arteria pulmonales, aorta ascendes, aortic arch, aorta descendes until bifurcation, arteriae coronariae, arteria carotis communis, arteria carotis externa and interna, arteriae cerebrales,

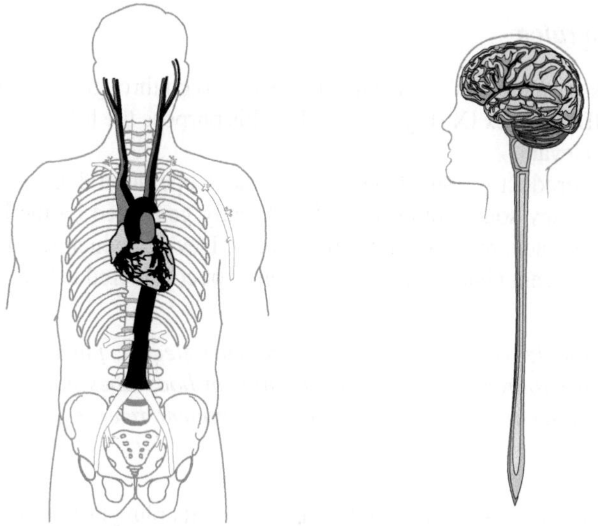

Figure 1-5: Central circulatory system (left) and central nervous system (right)

truncus brachiocephalicus, vena cordis, vena pulmonales, vena cava superior and inferior (e. g. catheter for intra-arterial blood pressure measurement, cardiac valve, stent).

5. ***special decisions*** directly define the conformity class of borderline products and may overrule the general classification scheme (e. g. devices used for contraception, blood bags, joint endoprostheses, breast implants), or assign them to medicinal products (e. g. X-ray contrast agents).

On the basis of these main product characteristics, for rough orientation a very simplified scheme for first-approach classification of medical devices is shown in Figure 1-6.

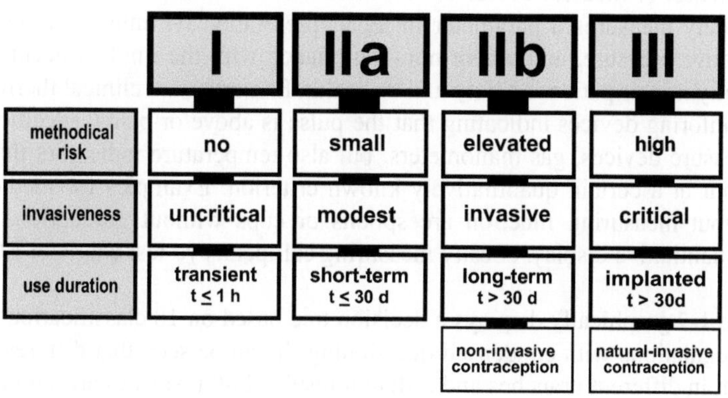

Figure 1-6: Simplified scheme for first-approach medical device classification into conformity class I, IIa, IIb and III

Classification rules

To reliably classify a medical device it is necessary to go through 18 classification rules laid down in MDD Annex IX (Figure 1-7). For this purpose the following further device properties are relevant:

Active: Any product is considered active, whose operation depends on an external energy source or any source other than that directly generated by the human body or gravity, and which acts by converting this energy. Devices which just transmit energy without any significant change or conversion are not considered active.

> *Remark: Examples of external energy sources are electric, pneumatic, hydraulic or radioactive sources. Energy from the human body does not make a device active unless such energy is stored for subsequent release (e.g. in a mainspring).*

Therefore, a syringe is not an active device because its plunger is activated by muscle force to deliver a substance to the patient. However, an implanted drug delivery device with a manually preloaded spring which subsequently enables delivery of the substance is considered an active device. Software is an active medical product because its operation depends on an external energy source. The electrode cable of an RF surgery device is not an active device because it passively transmits energy without change (the impact of the cable impedance can be neglected). However, the connected RF-cutting electrode is considered active because it concentrates energy to achieve the intended biological effect, and therefore it converts a current to high current density. Biosignal electrodes such as ECG-, EEG- and EOG-electrodes are not active because they are intended to pick up electric biosignals without change. Heating or cooling pads are not active if they just (passively) interact by their stored thermal energy while they would be considered active if they produce such energy by chemical reactions, endothermic or exothermic. Radioactive sources for tissue irradiation (e.g. brachytherapy-seeds) are considered active unless they are radiopharmaceutical substances (which are medicinal products).

Measuring: A medical device is considered to have a measurement function if it quantitatively measures a parameter in legal (preferably SI) units, or refers to such a quantitative measure, and where non-compliance with the implied accuracy could significantly impair patient's safety and/or health. Examples are clinical thermometers, pulse monitoring devices indicating that the pulse is above or below specified values, blood pressure devices, gas manometers, but also temperature indicators that change their colour at a certain quantitatively known criterion. Examples for medical products without measuring function are spoons or cups without graduation, droppers without quantitative display, obesity measuring callipers, eye-test charts or ECG paper etc.

Figure 1-7 graphically displays a decision tree based on 18 classification rules and specific regulations with condensed questioning. It can be seen that different features may result in different branches and different results. Full text rules are summarized in *Table 1-1* to *Table 1-4.* Different product aspects can result in different conformity classes. The highest identified class must be assigned to the device.

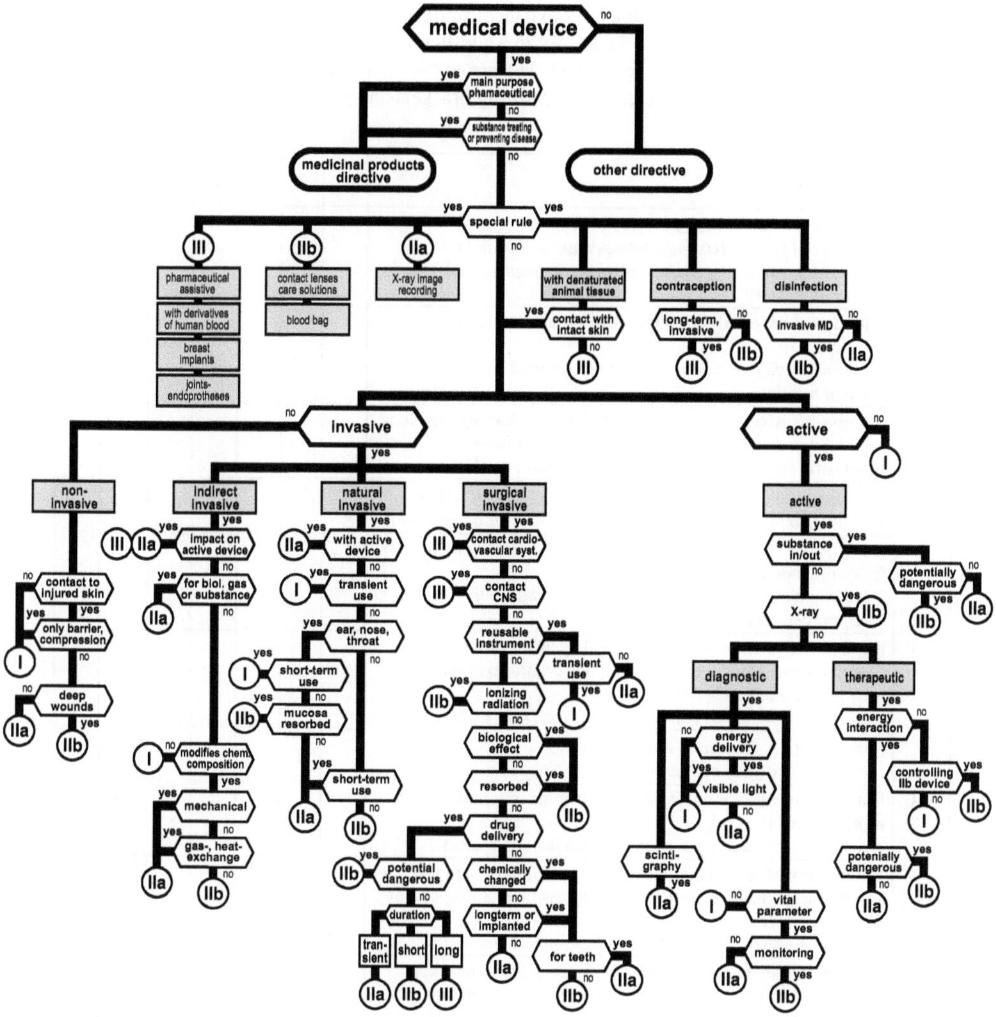

Figure 1-7: Decision tree based on 18 classification rules, potentially leading to several different classes for the same device. The highest class must be chosen.

Table 1-1: Classification rules for **non-invasive** medical devices (MDD, Annex IX). Grey fields signify the conformity class resulting if the answer to a question is "yes"; if a question does not apply, the "NA" (not applicable) -field must be marked. The arrow "⇨" indicates classification of the medical device (MD) into the conformity class written within brackets and indicated by the grey field in the associated right column.

No.	Classification rules	na	I	I_M	I_S	IIa	IIb	III
1	Is the product **non-invasive**? ⇨ **(I)** *e. g. hospital bed, walking aid, wheelchair, operation table, dental chair, corrective glasses, permanent magnets, orthopaedic leather appliances*							
2	Is it **non-invasive** and for **channelling** or **storing** liquids or gases? ⇨ **(I)** *e. g. gravity infusion administration set, syringes (without needle)*							
	Is it **non-invasive** and intended for **storing or channelling** blood, body liquids or tissues, or liquids or gases for infusion, administration or introduction **inside the body**? ⇨ **(IIa)**							
	Is it connected to an **active MD** class IIa or higher? ⇨ **(IIa)** *e. g. tubing or syringe for infusion pump, tubing for anaesthesia*							
	Is it intended for storing or channelling blood or body liquids or for storing organs or body tissue? ⇨ **(IIa)** *e. g. storage container for cornea, sperm, human embryos, transport containers for transplants*							
3	Is it **non-active** and for modifying biological or chemical composition of blood, other body liquids or infusion liquids … – by physical means? ⇨ **(IIa)** *e. g. blood filters, oxygenators, centrifuges, heat exchanger*							
	– by other (*e. g.* ***chemical***) means? ⇨ **(IIb)** *e. g. hemodialyzer, cell separators*							
4	Is it a **non-invasive** product intended to come into contact with **injured skin**? ⇨ **(IIa)** *e. g. devices to manage the micro-environment of wounds*							
	Is it **non-invasive**, coming into contact with injured skin and acting as a mechanical barrier, for compression or absorption of exudates? ⇨ **(I)** *e. g. wound dressings, absorbent pads, wound strips*							
	Is it **non-invasive**, and for contact with severely injured skin (with breached dermis and healing occurring by secondary intent? ⇨ **(IIb)** *e. g. dressings for chronic extensive ulcerated wounds or for severe decubitus wounds, dressing for temporarily substituting skin*							

Table 1-2: Classification rules for invasive medical devices (MDD, Annex IX)

No.	Classification rules	na	I	I_M	I_S	IIa	IIb	III
5	Is it a natural-invasive product and for connection to active MD class IIa or higher? ⇨ **(IIa)**, … *e. g. enteral feeding tubes, stomach drainage tube*							
	Is it a natural-invasive product without connection to active MD or for connection to MD class I and for **transient** use? ⇨ **(I)**, … or *e. g. dental aspirator tip, dental mirrors, examination gloves*							
	– for **short-term** use in oral, ear and nose cavity? ⇨ **(I)** *e. g. detachable dental prostheses, dressing for nose bleeding*							

No.	Classification rules	na	I	I_M	I_S	IIa	IIb	III
	– for short-term use outside oral, ear and nose cavities? ⇨ **(IIa)** *e. g. contact lenses, urinary catheters, tracheal tube, stent*							
	– for **long-term** use in oral, ear and nose cavities (without being resorbed)? ⇨ **(IIa)** *e. g. orthodontic wire, fixed dental prostheses, fissure sealants*							
	– for long-term use outside oral, ear and nose cavities? ⇨ **(IIb)** *e. g. urethral stents*							
6	Is it **surgical-invasive** for **transient** use? ⇨ **(IIa)** … *e. g. needles, single-use scalpel blades, drill bits*							
	– **except** … is it a **reusable** surgical instrument? ⇨ **(I)** *e. g. scalpels, saws, retractor forceps, excavators, chisels*							
	– is it specifically for direct contact with the **central nervous system** ⇨ **(III)**							
	– is it specifically for direct contact to the **heart** or **central circulatory** system for diagnosis, inspection or correction of a defect? ⇨ **(III)** *e. g. cardiovascular catheter, angioplastic balloon catheter*							
	– is it for supplying **ionizing radiation** energy ⇨ **(IIb)** *e. g. brachytherapy seed*							
	– is it for producing a biological effect? ⇨ **(IIb)**							
	– is it for wholly or mainly absorption? ⇨ **(IIb)**							
	– is it for administering medicines via a delivery system in a potentially hazardous way? ⇨ **(IIb)** *e. g. insulin-pen for self-administration*							
7	Is it **surgical invasive** for **short-term** use? ⇨ **(IIa)** … *e. g. clamps, infusion cannulae, temporary filling material*							
	– **except** … is it particularly for direct contact to the heart or central circulatory system for diagnosis, inspection or correction of a defect? ⇨ **(III)** *e. g. cardiovascular catheter, temporary pacemaker lead, carotid artery shunt*							
	– is it particularly for direct contact with the **central nervous system?** ⇨**(III)** *e. g. neurologic catheter, cortical electrodes*							
	– is it for releasing ionizing radiation? ⇨ **(IIb)** *e. g. brachytherapy device*							
	– is it for producing a biological effect? ⇨ **(III)** *e. g. biologic adhesive*							
	– is it for wholly or mainly absorption? ⇨ **(III)** e. g. absorbable sutures							
	– is it for undergoing chemical change (except devices placed in teeth or administering medicines)? ⇨ **(IIb)**							
8	Is it **surgical invasive** for **long-term** use or implantation? ⇨ **(IIb)** *e. g. shunts, stents, nails, plates, intra-ocular lens, infusion ports*							
	– except … is it to be placed in teeth? ⇨ **(IIa)** *e. g. bridges, crowns, dental filling material, ceramic*							
	– is it for direct contact to the heart or central circulatory system? ⇨ **(III)** *e. g. prosthetic heart valve, aneurism clip, vascular prostheses and stents*							
	– is it for direct contact with the central nervous system? ⇨ **(III)** *e. g. CNS electrodes, spinal stents*							
	– is it for producing a biological effect? ⇨ **(III)** *e. g. adhesives*							
	– is it for wholly or mainly absorption? ⇨ **(III)** *e. g. absorbable sutures*							

No.	Classification rules	na	I	I_M	I_S	IIa	IIb	III
	– is it for undergoing long-term chemical change (except devices placed in teeth or administering medicines)? ⇨ **(III)** *e. g. bone cement would change too rapidly (already during placement)*							
	– is it for administering medicine? ⇨ **(III)** *e. g. rechargeable non-active drug delivery systems*							

Table 1-3: Classification rules for active medical devices (MDD, Annex IX)

No.	Classification rules	na	I	I_M	I_S	IIa	IIb	III
12	Is it an **active** product? ⇨ **(I)** *e. g. surgical microscope, hospital bed, patient hoist, walking aid, wheelchair, stretcher, dental patient chair, thermography device, dental curing light, recording, devices for processing or viewing diagnostic images …*							
9	– **except** … is it for **therapeutic** administration or exchanging **energy** … a) **without** potential hazards? ⇨ **(IIa)** *e. g. ergometer, muscle stimulator, electric acupuncture, ultrasonic therapy*							
	b) **with** potential hazards? ⇨ **(IIb)** *e. g. lung ventilators, baby incubators, warming blanket, blood warmers, RF surgery devices (including electrodes), defibrillator, surgical lasers, surgical ultrasound devices, X-ray therapy devices, afterloading devices*							
	– is it for controlling, monitoring or directly influencing performance of an active therapeutic device class IIb? ⇨ **(IIb)** *e. g. therapy planning software, afterloading control devices*							
10	Is it an active product for **diagnosis** and … – for supplying energy which is absorbed by the human body (except for illumination with visible light)? ⇨ **(IIa)** *e. g. MRI, diagnostic ultrasound, evoked response stimulator*							
	– for imaging in-vivo distributions of **radiopharmaceuticals**? ⇨ **(IIa)** *e. g. Gamma-cameras, SPECT, PET*							
	– for directly diagnosing or monitoring vital physiological processes … a) without indicating acute danger? ⇨ **(IIa)** *e. g. ECG recorder, EEG recorder, electronic thermometers, electronic blood pressure measurement devices, electronic stethoscopes*							
	b) for indicating acute **danger**? ⇨ **(IIb)** *e. g. monitors for ECG, respiration, blood pressure, blood gases*							
	Is it an active product emitting **ionizing radiation** for diagnosis or guiding surgical interventions? ⇨ **(IIb)** *e. g. diagnostic X-ray devices*							
	– or is it for controlling, monitoring or influencing the performance of such devices? ⇨ **(IIb)** *e. g. dosimeter*							
11	Is it an active product for administering and/or removing **substances** (e. g. medicines, body liquids) to or from the body? ⇨ **(IIa)** *e. g. suction devices, feeding pumps, jet injectors, nebulizers*							

No.	Classification rules	na	I	I_M	I_S	IIa	IIb	III
	– except … is this done in a potentially **hazardous** manner (in regard to substances, site of body, mode of application)? ⇨ **(IIb)** *e. g. infusion pumps, ventilators, anaesthesia machines, dialysis equipment, blood pumps for heart-lung machines, hyperbaric chambers, medical gas mixers, drug nebulizers, critical care moisture exchangers*							

Table 1-4: Special classification rules for medical devices (MDD, Annex IX)
Grey fields signify the resulting conformity class if the answer to a question is "yes," if a question does not apply, the field "NA" (not applicable) must be marked

No.	Special classification rules	na	I	I_M	I_S	IIa	IIb	III
13	Is it incorporating an assisting medicinal product? ⇨ **(III)** *e. g. heparin-coated catheter, antibiotic bone cement, spermicide-coated condoms, ophthalmic irrigation solution with metabolism-supporting component, dressing incorporating an ancillary antimicrobial agent*							
	Is it incorporating as an integral part human blood derivates? ⇨ **(III)**							
14	Is it for contraception or prevention of the transmission of sexually transmitted diseases? ⇨ **(IIb)** *e. g. condoms, diaphragms*							
	except … is it implantable or long-term invasive? ⇨ **(III)** *e. g. contraceptive intrauterine devices (if the primary purpose is releasing progestogens, IUDs are considered medicinal products)*							
15	Is it for disinfecting medical devices? ⇨ **(IIa)** … *e. g. endoscope-disinfectants, washer-disinfectors, sterilizers (just cleaning means are not included)*							
	Is it specifically for disinfecting invasive devices? ⇨ **(IIb)**							
	except …Is it for disinfecting, cleaning, rinsing, or hydrating contact lenses? ⇨ **(IIb)** *contact lens solutions, comfort solutions*							
16	Is it for recording X-ray diagnostic images? ⇨ **(IIa)** *(this does not include media for subsequent reproduction)*							
17	Is it utilizing **animal tissues** or its non-viable derivatives and not contacting intact skin only? ⇨ **(III)** *e. g. biologic heart valves, catgut sutures, implants and dressings of collagen (excluded are orthopaedic leather appliances, milk, silk, beeswax, hair or lanolin)*							
18	Is it a blood bag for storing purposes with or without coatings of anticoagulants? ⇨ **(IIb)** *(if function goes beyond sole storing or substances for preservation other than anticoagulants are included, other rules – e. g. Rule 13 – apply).*							
	Is it a class I device with measuring function? ⇨ **(I_M)**							
	Is it a class I product marketed sterile? ⇨ **(I_S)**							
			I	I_M	I_S	IIa	IIb	III
Final score (the highest class is applicable):								

1.4.4 What is the meaning of the CE-mark?

Today, CE-marks are omnipresent in Europe. However, only few persons really know much about their meaning. Customers are confused by the inflationary use, salesmen claim they proved that products meet high-level European requirements, and manufacturers ask who could "grant" them the CE-mark. Often, CE-marks are advertised as quality marks or safety marks, and it is claimed they would indicate that a product was "CE-certified".

But is the CE-mark indeed an indicator of tested and approved safety? Does it signal governmental market approval or verified outstanding quality? ... or is it simply an assertion of the manufacturer that he had met the rules – as credible or not, just as the manufacturer's own reputation. But if CE-marking indicated compliance with requirements another question arises, namely, which ones? CE-marks can be found on a manifold of products, from tooth brushes, teddy bears and computers to heart – lung machines! Confusion is great.

The initial intention of CE-marking was (just) to overcome trade barriers and to signal authorities that so-marked products should be allowed to be marketed without further hindrance. This is the reason why so diverse products carry the same mark in spite of their different type, function and composition. Even reference to requirements of European Directives is of little help if someone is not aware, which of the many directives had been applied and what manufacturers had to fulfil to CE-mark their products.

Even if it is clarified that on a medical device a CE-mark indeed refers to the medical devices directive and not to the low-voltage directive, the cosmetics directive, or to a directive that just covers a particular aspect such as the directive of electromagnetic compatibility or the pressure vessel directive (Figure 1-1) – even if it were affixed and all legal requirements were met, the CE-mark may still have a variety of meanings such as

- an untested self-declaration of the manufacturer (for conformity class I products);
- an indication that the product has been manufactured reproducibly – which must not be misinterpreted with outstanding quality. "Constant" quality may be good – or less good. Third party testing is restricted to manufacturing according to the (untested) technical file generated by the manufacturer (for conformity class IIa products);
- an indication that the design of the product and the manufacturing process has been evaluated, tested and certified by an independent third party (European Notified Body) (for products of conformity class IIb and class III).

Therefore, CE-marks can only be clearly interpreted with additional knowledge of the requirements applicable to medical devices, their conformity class and which modules of conformity assessment the manufacturer has chosen.

1.4.5 How do devices get a CE-mark?

CE-marking is required for putting a medical device on the market, but does any medical device need to be CE-marked? There are exemptions. CE-marking is ***not*** required for:

- medical devices ***not*** (yet) put on the market. This is the case if products are just presented to clients such as at fairs, provided their performance is not demonstrated on human volunteers. However, if devices are stored ready for delivery to clients such as in a manufacturer's warehouse they are considered to be already on the market and must bear a CE-mark.
- medical ***systems*** (e. g. functionally connected components) and medical procedure packs (e. g. procedure-specific compositions of products) consisting of already CE-marked components that continue to be used as intended by their manufacturer. However, if such a system or pack contains at least one component not yet CE-marked or is used beyond its intended specifications (e. g. a power supply unit loaded more than its rated value), the whole system or pack needs to be reassessed and must be CE-marked as a product in its own right.
- ***custom-made*** devices. These are devices manufactured for a particular named individual. They are not made available for general use and, hence, are not considered as being on the market. Examples are dental prostheses fitted to an individual (however, not industrially produced teeth), artificial legs, reading glasses (however, not spectacle frames or glass blanks). Devices of general use specially designed according to the wishes of a health care unit are not considered custom-made and need CE-marking.
- ***in-house production***. Medical devices produced at own premises for own needs and not intended to be handed out to third parties are not considered as on the market and, hence, do not need CE-marking. However, they must be conformity assessed and meet the essential requirements, anyway. The term "in-house" refers to single autonomous functional units such as a particular hospital. However, devices produced in one hospital intended for use also in another hospital of the same legal organization need to be CE-marked (with potential third-party assessment and certification).
- ***second-hand.*** Already used devices put again on the market for second-hand use, even after repair do not need reassessment and re-CE-marking. However, devices that have been refurbished to make them "like new," must be reassessed and re-CE-marked. Refurbishment could also include single-use devices to make them reusable (against the intended use as defined by the initial manufacturer). However, now the refurbisher becomes the new manufacturer with all rights and obligations.

  Remark: In contrast to repair which just aims at correcting defects but leaves the product in an age-appropriate condition, refurbishment is understood as a procedure of completely overhauling a product including replacing parts potentially degraded by age to make it "like new."

- ***clinical testing***. Products required for clinical testing obviously have not already completely passed conformity assessment and, hence, are not allowed to bear a CE-mark. However, instead, they may be used for clinical testing (provided the appropriate procedure is followed, see Chap. 3.2) but must be labelled as "for clinical testing".

Remark: Clinical studies must be approved by an Ethics Committee and permission must be applied for from the competent authority. A clinical study can only be started after permission (or non-prohibition) by the competent authority.

Requirements for CE-marking

A CE-mark is neither "granted" nor "awarded" to the manufacturer by anybody. It is the manufacturer himself who affixes it by his personal responsibility provided the legal and safety requirements are met. This includes assessment and documentation conformity with the essential requirements. However, adequate design is not sufficient. In addition, it is required that devices are reliably manufactured according to the technical file. Conformity self-assessment by the manufacturer might not be sufficient in any case. Depending on the inherent risk of a device involvement of a recognized third party (a European Notified Body) might be mandatory. The extent of mandatory third party involvement is different depending on medical device's conformity class (Figure 1-8):

Conformity class I: Products with no or only small risk potential are exempted from third party testing. In that case the manufacturer is allowed to assess conformity on his own; afterwards he elaborates the technical file and the conformity assess-

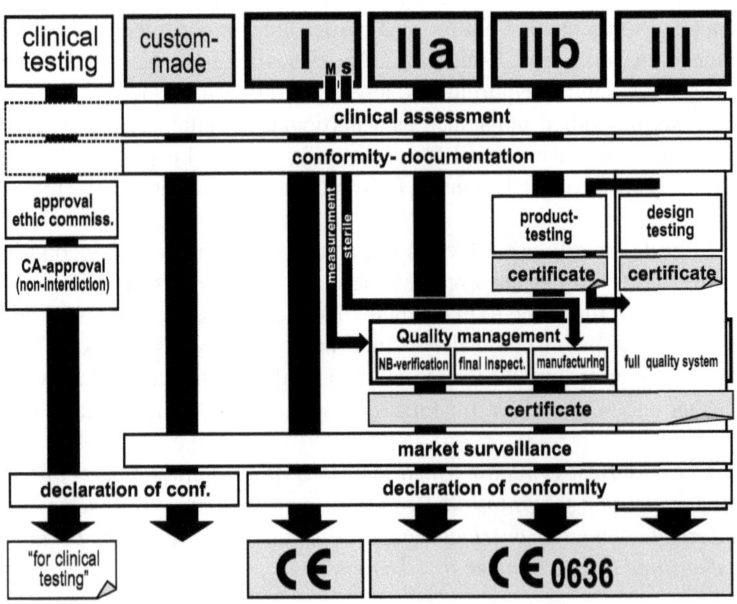

Figure 1-8: Paths to CE- marking and inclusion of European Notified Bodies for conformity assessment and certification (indicated by the "certificate" boxes)

ment file, draws up a legally binding declaration of conformity and affixes the CE-mark on the product and its packaging (without adding any notified body identification number).

> *Remark: If a medical device bears a CE-mark without an identification number it signals at first glance that it belongs to conformity class I – or that it is inappropriately classified and, hence, illegally marketed.*

Products with a ***measuring function*** or ***sterile*** products require certified quality assurance at least for these features, even if they belong to class I. Therefore, conformity class I is subdivided into subclass I_M (class I devices with measuring function) and subclass I_S (sterile devices class I). However, modules appropriate for sterilization quality management are restricted to at least to a quality management system according to Annex V MDD which involves the whole sterilization process. Restriction to final inspection (Annex IV MDD) is not sufficient.

Conformity class IIa: Devices with an inherent risk potential require quality-assured manufacturing (according to the technical file of the manufacturer) based on one out of four optional modules. While product conformity assessment may still be performed by himself the manufacturer must implement a quality management system which requires auditing and certification by a European Notified Body. A valid QM certificate is the precondition that manufacturers can issue the declaration of conformity and affix the CE-mark on the produced medical devices and their packaging. Involvement of a European Notified Body is indicated by adding its 4-digit identification number to the CE-mark (e. g. **CE**0636 for the European Notified Body PMG of Graz University of Technology).

> *Remark: Quality management systems for medical devices require certification by a European Notified Body rather than by a general quality assessment institution, even if accredited.*

Conformity class IIb: Devices that are characterized by an elevated risk potential require also third-party conformity assessment of the product including safety, performance and usability (EC-type testing). Basically, EC-type testing does not mean that manufacturers would not be allowed to further change and/or develop their device. This still remains possible and usually is the common case. However, relevant changes must be assessed and released by the European Notified Body prior to implementation in manufacturing.

After passing EC-type testing, devices additionally require quality-assured manufacturing according to the type-test approved technical file. Manufacturers may still choose one out of the four QM modules (Figure 1-8). The chosen quality management system needs auditing to determine whether it meets the related requirements and issuing of a QM certificate based on regular surveillance audits. Therefore, conformity class IIb devices require two certificates (EC-type testing and QM system auditing). These certificates are the precondition that manufacturers can issue the declaration of conformity and affix the CE-mark with the accompanying 4-digit notified body identification number.

Remark: The two required certificates may be issued by two different European notified bodies. The CE-mark is added with only one identification number which is that of the QM-certifying body.

Conformity class III: Devices with high risk potential must be designed, manufactured and marketed under a full quality management system which must be assessed, certified and regularly audited by a European Notified Body. Manufacturers may perform own EC-type testing of their devices provided they have an own test department which is operated within the QM system and meets all requirements of accredited test bodies in regard to organization, competence and equipment. As an alternative, manufacturers may outsource EC-type testing to a European Notified Body. If EC-type testing is performed in-house, in addition to QM system auditing the design of the product needs to be examined based on the documentation, and in addition to the QM system certificate a design examination certificate is issued. Therefore, conformity class III devices also require two certificates to allow manufacturers to issue the declaration of conformity and affix the CE-mark with the accompanying 4-digit notified body identification number.

Certificates are valid for a maximum of 5 years. If necessary they may be extended on application (and after potential supplementary testing) for further periods of a maximal length of 5 years.

1.5 Administrative obligations

Medical device manufacturers must either have a registered place of business in one of the EU member states or designate an authorized representative in the EU. The manufacturer or his authorized representative must meet the following requirements:

- the manufacturer or his authorized representative must register himself and the marketed medical device(s). This is done unbureaucratically just by visiting the homepage of the competent authority[1] responsible for the site of the manufacturer (or his authorized representative). In the course of interactive registration the medical device must be assigned to one of the available device categories and characterized by a numerical code[2]. Menu-guided registration is free of charge. Registration results in registration numbers for the manufacturer and the various registered devices. These numbers are essential identification tools in the European Medical Devices Register. They are cited in the certificates and must be indicated in vigilance reports.
- the manufacturer must issue a statement of conformity and keep it with the records to forward it to the competent authority, if requested.

---

[1] In Austria via http://medizinprodukte.oebig.at, in Germany via homepages of provinces. Information can be found at http://www.dimdi.de/medizinprodukte/zuständigeStellen.

[2] Remark: For the time being the European Medical Devices Register uses the code of the Universal Medical Device Nomenclature Systems (UMDNS). In the future this coding system will be replaced by the Global Medical Device Nomenclature (GMDN) System, which contains 12 main categories with appr. 7,000 items und over 10,000 synonyms.

- the manufacturer or his authorized representative must be able to provide competent authorities with the technical file of the device on request in due time.

  Remark: Since technical files may also contain sensitive know-how of the manufacturer which he may not be willing to disclose to his authorized representative, it is accepted if transfer of the technical file is done directly from manufacturer to competent authorities (which are obliged to confidentiality). In cases where manufacturers are located outside the EU the authorized representative is obliged to make the technical file available. Therefore, representatives are recommended to conclude a contract with the manufacturer clearly regulating this issue.

- the manufacturer must ***store records*** and make them available on request of a competent authority until at least 5 years after the last device has been produced. For implanted medical devices the period is extended to 15 years after production.
- the manufacturer or his authorized representative must ***notify*** the competent authority of any of the following events:
  - ***systematic call-backs of devices*** caused by technical or medical reasons.
  - ***severe** (unexpected) **incidents*** that have already occurred or almost occurred caused by a device due to malfunction or deterioration in the characteristics and/or performance, inadequacy in the labelling or instructions for use, or degraded quality. Severe incidents are such that led to death or serious deterioration of health. As an example, severe burns due to electric muscle stimulation must be reported since they should not be expected while death of a patient following defibrillation according to the instructions for use must not be notified since a 100% success rate cannot be expected for this application.

  Reporting must be done within a limited period after the occurrence (or the awareness) of the event. The notification tolerance period is 10 days after an incident and 30 days after a near-incident starting from the event or information about the event. Since notification must be done by several groups such as physicians, operators and clinical engineers, competent authorities are able to verify whether all involved parties have fulfilled their obligation to report.

> **Infection on sale**
>
> **Massachusetts:** The manufacturer of RF surgery electrodes had to recall several batches of electrodes because non-sterile devices had erroneously been marked as sterile. This failure could have led to infections with major health risks including collapse of organs or death.

1.6 Organizational obligations

The manufacturer or his authorized representative must implement and maintain a risk management process to monitor risks and examine assumptions made for risk assessment (Chap. 2.2). This includes maintaining a market surveillance system allowing awareness of unintended events and incidences, and if necessary, taking appropriate actions such as reassessing risks based on updated failure probability or implementing additional risk-reducing measures (Chap. 2.2.4).

- the manufacturer and/or his distributer must have sufficiently qualified personnel to competently inform clients about the device and professionally train users.

  Remark: Manufacturers must have a sufficiently trained "medical devices consultant."

- the manufacturer and/or his distributer must implement a market surveillance system to actively collect data allowing assessment of use experience and identification of unintended events and incidences.
- the manufacturer and/or his distributer must have sufficiently qualified personnel to assess data from market surveillance and to competently decide upon additional risk control activities.

  Remark: Manufacturers must have a sufficiently trained "medical devices safety officer."

In addition, depending on the kind of product, authorities may oblige manufacturers:

- to keep product records to enable call backs and care for affected patients in case of product failures;
- to implement additional quality management activities.

1.7 Legal obligations

The European New Approach grants manufacturers far-reaching decision power. It is the manufacturer who assigns his product to a conformity class and subsequently selects the path leading to CE-marking; it is the manufacturer who assesses conformity, and finally it is the manufacturer who affixes the CE-mark with sole responsibility. Since CE-market supervision is still being implemented, the question arises what is the manufacturer's motivation to work within the rules and to follow the demanding legal path to CE-marking, when ignoring the rules and subsequent savings in time, costs and effort are so tempting? Fines do not have much of a deterrent effect and any potential loss of image might be sweetened by profit made in the meantime.

A major motivation to stick to the rules is liability regulations (see Sect. 1.7.5). The consequences of a faulty product such as callbacks or liability for subsequent damage could produce costs of a magnitude which could be high enough to endanger the economic existence of a manufacturer. Experience shows that manufacturers are less careful with alleged low-risk products while risk management is taken more seriously for high-risk products in awareness of their elevated risk potential. This explains why severe and even deadly accidents happen with simple hospital beds because of underestimated existence and/or frequency of hazards such as fire due to short-circuits, electric shocks, contusions, strangulations or falls. Therefore, manufacturers should be motivated to follow the rules and to produce reliable products that meet legal requirements simply because of their instinct for self preservation.

1.7.1 Declaration of conformity

After appropriate conformity assessment and documentation manufacturers are obliged to write a clear and legally binding declaration that their products meet the applicable legal requirements. Together with other required documentation the declaration of conformity must be made available to authorities if necessary; usually it is also included in the instructions for use.

> **Declaration of conformity**
>
> We herewith declare, that our product including accessories meets all national and international CE-standards, according to the medical devices ordinance, class 2, safety class 3b.

However, frequently such declarations are deficient. For example (see text box), if a manufacturer declares conformity "with all national and international requirements," or refers to "CE-standards" and classifies his device "class 2, safety class 3b," he clearly demonstrates his ignorance of regulations in general (the term CE-standard is a non-existent nomenclature) and in particular which regulations were indeed applicable to his product and even of the intrinsic features of his devices: a "class 2" is non-existent – it would either be safety class II or conformity class IIa or IIb; the same applies to the mentioned "safety class 3b," which is also non-existent because there is only a safety class III – which medical devices are not allowed to use (Chap. 8.5)– or a conformity class III.

While it is left open to the manufacturer how to design and formulate his declaration of conformity, the minimal contents of such a declaration are defined. An example would be:

the manufacturer
(name and address)
declares in sole responsibility,
(without restrictions or reference to subcontractors or suppliers),
that his product
(identified by the name of type or product family);
meets the essential requirements of the medical devices law *(of a EU member state),*
(the reference to a national law allows filing a lawsuit – which would not be possible with the sole EU directive);
and the European Medical Devices Directive 93/42/EC
and that it belongs to conformity class …,
and was manufactured according to the following regulations
(the listing of applied standards or the chosen deviating equivalent solutions);
and was put on the market based on the following conformity modules which is certified by the following certificates … .
(indication of valid (!) certificates and the issuing European Notified Body).
legally binding signature (of the authorized person)

1.7.2 Confidence

Even if it is hard to believe after having a glance at the newspapers or watching TV news: A fundamental basis of our (social) life is mutual trust. This trust extends to all situations: we trust in the safe construction and maintenance of the elevator we use, in the beneficial composition, hygienic production and storage of food we eat, we trust in the correct behaviour of other participants in road traffic – and of course in particular and notably we have to trust in health care, in competent diagnosis and efficient therapy … and in the safety and reliability of medical devices.

1.7.3 Carefulness

Legislation has added to the principle of justified trust in others the obligation that everybody has to do everything with a reasonable amount of care[3].

A cleaner who in spite of adequate training waxes the costly electric conducting floor of an operating theatre with insolating wax (therefore compromising expensive measures against dangerous electrostatic charging) is violating due diligence. The same goes for a technician who after recurrent device testing did not perform final function testing and consequently did not become aware that he had damaged the device during insulation impedance measurement or by destroying an electronic component by electrostatic discharging during internal visual inspection. A surgeon who caused severe burns to a patient because of careless attachment of the RF surgery neutral electrode also violates due diligence.

! Due diligence obligates everybody to carefulness

However, when demanding diligence lawmakers do not only restrict themselves to sole carefulness. An untrained service engineer who carefully adjusts the mirrors inside a class 4 laser device and leaves his fingerprints on them cannot claim not to be responsible for subsequent costly thermal damage caused by the intensive light which subsequently ignited the deposited fat. Lack of knowledge is not accepted as an excuse. The same applies to an electrician who performs an electric installation in a surgeon's practice in the same way as he was used to doing in dwellings and thereby ignoring the specific installation standards applicable to locations for medical use – or to a technician who clamps soldered lacings in the same way as he was taught at school several decades ago. The reason why lack of knowledge is not accepted as an excuse is that in addition to carefulness, due diligence comprises also the obligation of keeping one's level of knowledge up to date and of restricting one's own activities to those tasks only for which one's qualifications are sufficient.

Expensive employee

Negligent service technician

Illinois: A manufacturer had to call back six-channel infusion pumps maintained during May 22 until August 7, 2007 because in contrast to his pretention a service technician had not performed the intended software-upgrades.

[3] General Civil Code.

Lack of knowledge is no excuse

However, similar to other laws updating one's level of knowledge is an own responsibility rather than the obligation of others. The excuse "that nobody told me that new regulations existed" is not acceptable. Everybody must actively keep their level of knowledge up to date on their own.

The obligation for sufficient qualification leads to the consequence that everybody is only allowed to perform tasks for which they have sufficient qualification. Violating this requirement by performing tasks which go beyond one's own skills is classified as careless acceptance. Therefore, in our time of rapid growth of knowledge continuous education has become an ongoing challenge.

Carefulness requires conscientiousness and competence

Although there is no lack of car repair shops, car drivers are required to have minimum technical knowledge (at least at the time of the driver's examination). Likewise, medical staff is requested to have basic knowledge of safe application and specific risks associated with medical devices.

Lethal dialysis

Patient bled to death

Graz: During dialysis a patient lost more than 1 l of blood because a tube had loosened from the catheter. Three days later he was dead. A charge was pressed against the hospital. However, it was not the hospital that was convicted but the nurse who had affixed the tube as usual. Two physicians had been arguing that the fixing of tubes was not their job. Although they hardly could explain the function and risks of the device the nurse was sentenced to a high fine; the physicians and hospital were just criticized.

This became evident in the following example. In the USA during a circumcision a surgeon caused severe burns to a patient because he used RF surgery instead of a scalpel. In the following lawsuit the patient was adjudged a high compensation. Who do you think had to pay this? It was not the physician but the manufacturer of the device. The justification was that the instructions for use did not contain a warning not to apply this method to such an indication.

In contrast to the situation in the USA, in Europe the obligation of occupational competence has far-reaching consequences. The same case would have been considered medical malpractice since it is requested that users of medical devices are aware of their intended use. Therefore, by defining the intended use manufacturers limit their product liability. Consequently, if someone dries his pet in the microwave oven or cuts hedges with his lawn mover he does this (in Europe) on his own risk and responsibility. In case of an accident he/she may not claim for compensation but even would face accusations of negligence.

Remark: It is only the manufacturer who defines the intended use of a device. It is laid down in the instructions for use of his product. Therefore, it is the manufacturer who defines whether an infrared radiator may also be used by lay people (without special knowledge or insight into risk potentials) at their homes (without connection to a reliable electric installation) instead of being used by medical staff only (with special training) in a medical environment (with reliable installation). It is also the manufacturer who defines under which conditions his product may be used e. g. whether it is designed to be used also in an explosive atmosphere.

1.7.4 Warranty

Warranty concerns the right of a customer to a fault-free product (European Directive 1999/44/EC /12/) while liability refers to consequential damage caused by a faulty product. Warranty means that a customer who purchased a product for money or money's worth is entitled to get it in a generally expectably fault-free condition, and that it indeed offers the advertised performance. If this is not the case, the salesman (and not the manufacturer or distributer) is obligated to warranty the product.

If in spite of the deficiency the intended use is still possible the customer has (only) the right to have the fault fixed. Instead, he might negotiate a price reduction. However, the *right* does not exist to exchange the product for another one which is fault-free. Only if remedy of the fault is not possible and if it is impossible to use the product as intended the client has the right to step back from the purchase.

The right to warranty starts from the date of purchase and extends until a deadline of 2 years for movable goods and 3 years for immovable goods. Within the first 6 months the salesman has the burden of proof /12/. Within this period, he has to prove that the product was free of faults when handed over. Afterwards, the burden of proof is shifted to the customer.

Reduction of warranty is not permitted for new products. The only exceptions are second-hand products provided they are sold privately to private persons. In that case it is allowed to reduce or even exclude warranty.

Remark: In contrast to legally regulated warranty, guarantee is a voluntarily offered obligation which might be bound to additional requirements. However, it must not be less than legal warranty.

1.7.5 Product liability

Liability for defective products is regulated by the European Directive 85/374/EG and the related national laws /56/. According to these regulations the manufacturer or, if not known, the supplier, shall be liable for damage caused by a defect in his product (which includes software and energy). Liability presumes damage during intended or reasonably expectable use. It does not apply to enterprisers using a product mainly in their own premises. Liability comprises consequential damage to life and health as well as damage to or destruction of any item or property other than the product itself. However, it does not include financial loss such as reduction of revenue or sales.

Explosive irradiation

Berlin: Twelf minutes after starting using an infrared irradiation lamp exploded and caused severe damage to user's face. Because of product liability the manufacturer was fined to financial compensation for damage and pain. His appeal was dismissed. He had claimed that explosion of a lamp at the end of its lifetime would not be unusual and therefore must not be seen as a fault.

Liability expires 10 years after putting the product into circulation. A product is considered defective if (when handed over – not when the damage occurs) it does not provide the safety which a person is entitled to expect. However, liability cannot be demanded for the sole reason that subsequently a better product has been put into circulation. Legal claims prescribe after a limitation period of 3 years from the day on which the plaintiff became (or should have reasonably become) aware of the damage.

Liability comprises damage only above € 500 but without further limitation. It comprises

- damage to property;
- personal injury (without limitation or retention);
  - costs of cure and care;
  - compensation for pain and suffering;
  - omitted alimony (after death of the liable person);
  - loss of earnings;
  - loss of advancement because of defacement.

Manufacturers are liable independent of negligence. However, they shall not be liable if *they* are able to prove (shift of burden of proof) that:

- the product was not defective when handed over;
- the damage is not causally related to a product failure;
- the product was not put into circulation (e. g. an exhibit that had been used prior to clearance);
- the product was not handed over for money or money's worth;
- the product was not deficient because it was in compliance with the initially (!) accepted state of science and technology even if in the meantime shortcomings in standards or scientific assumptions had been identified;
- the defect is due to compliance with legal requirements;
- the product was not used as intended.

The proof of freedom from defects (at the time when the product was handed over to the customer) must comprise design, construction, manufacturing, distribution and storage. It can be given by a type-test certificate, a manufacturing quality management certificate and proofs of adequate package, transport and storage.

1.8 Opportunities und pitfalls

A manufacturer has many possibilities to decide upon the complexity, duration and costs of product market approval. The strategic decisions he takes influences the conformity class and the path to conformity approval and CE-marking and may offer opportunities but could as well contain pitfalls.

The most important decisions which influence assignment of a product to a conformity class and, consequently, the consequences in regard to complexity and expenses of the conformity assessment procedure are related to the following aspects (Figure 1-9)

- the ***intended function***. From this it follows whether a product becomes a medical or a non-medical device (e. g. if intended for medical, wellness or universal use). The intended use might determine whether a product may be considered hazardous or harmless – which leads to quite different conformity classes. This applies to
  - ***intended purpose***, e. g. for therapy (medical product), for well-being (wellness product), for appearance (cosmetic product or general product). As an example, a UV lamp could be intended for treating dermatosis (medical device), for tanning (cosmetic device), for illumination of discos (general electronic device), or for hardening glued joints (industrial device).
  - ***health relevance***, e. g. uncritical (e. g. ECG recording) or life-saving (e. g. ECG monitoring); uncritical (e. g. electrotherapeutic muscle stimulator) or life-sustaining (e. g. phrenic stimulator for paralyzed patients);
  - ***output***, e. g. delivery of uncritical or hazardous amounts of electric current, voltage, radiation, substances, pressure or heat;

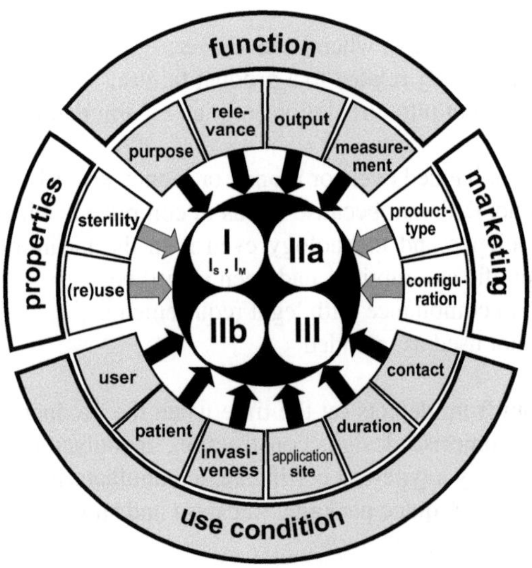

Figure 1-9: Options for strategic decisions influencing medical product assignment to conformity classes

  - ***measurement,*** e. g. relative indication or quantitative values, critical or uncritical parameters.
- ***use conditions***
  - ***intended users***, e. g. trained medical staff or lay people;
  - ***patient's condition***, e. g. conscious, unconscious, sane or insane, device-dependent or not;
  - ***invasiveness***, e. g. non-invasive, natural-invasive, surgical-invasive;
  - ***site of application***, e. g. suction devices could be intended for use inside the mouth (class I), inside surgical wounds (class IIa) or inside the lung (class IIb);
  - ***kind of contact***, e. g. contactless, contact with clothing, intact or wounded skin, surgical wounds, blood circulatory system, heart or central nervous system;
  - ***duration of use***, e. g. transient, short-term, long-term, permanent.
- ***product characteristics***
  - ***condition***, e. g. non-sterile or sterile;
  - ***usability***, e. g. single-use or reusable;
  - ***expected service life***, e. g. unlimited use or time-limited use until a given expiry date;
- ***product configuration***, e. g. components marketed as a medical system with a conformity class according to the worst-case component or as individually CE-marked devices with individual conformity classes.

Apart from deciding upon the conformity class manufacturers may also make a choice between different modules for access to CE-marking (Figure 1-8). Among others, he may decide upon certification costs which may be dependent or independent of production:

- flexible piece-related costs that are easy to calculate follow from outsourcing quality management (final inspection) to a European Notified Body. In that case costs arise only if a device is actually produced and sold;
- almost production-independent fixed costs follow from implementing and maintaining an own quality management system. Except for class III devices where full quality management is obligatory, manufacturers can choose among quality management systems restricted to final inspection, comprise the whole manufacturing process or involve all product-related activities.

After these strategic decisions the manufacturer can proceed on the path to product marketing. The required steps comprise registration of the company and the marketed product types, selection of conformity modules, implementing the risk management process, performing product design and conformity assessment, documentation, quality management, certification (if necessary), writing the declaration of conformity, CE-marking and market surveillance (Figure 1-10).

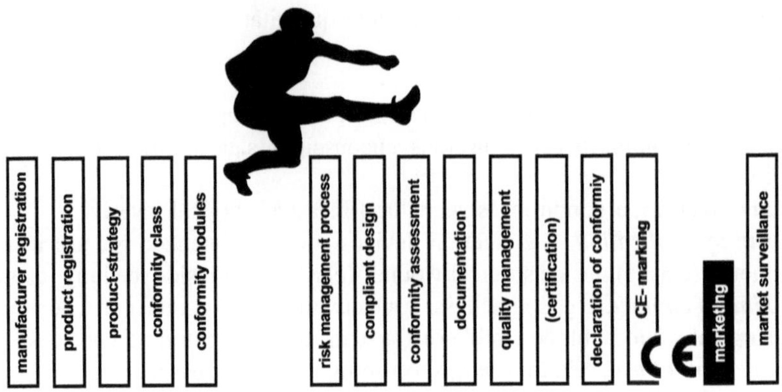

Figure 1-10: Manufacturer's steeplechase to product marketing

2 How safe is safe enough?

Did anything go wrong today? Did you forget something or miss an appointment because you were stuck in a traffic jam? Did you only just miss a bus or did your shoelace tear at a time when you were in a hurry? Such mishaps make us aware that we are continuously exposed to risks. Fortunately, usually we succeed in handling them, of course thanks to the fact that we have developed individual risk management strategies.

2.1 Risk

Initially, safety (Latin: *sine cure*) was understood as being free of sorrows, being able to confide in someone without any concern. In that sense, something would be "safe" if it was free of any hazard. However, a hazard (caused by a hazard source) does not already by itself make harm unavoidable. In fact, usually harm is the final consequence of a series of events originating from a hazard and triggered by some incidence (Figure 2-1).

The presence of a hazard just leads to the *possibility* that a hazardous situation might occur which after a triggering event with some probability could lead to harm. However, usually harm does not already follow from one single hazardous situation. As a rule, there usually exist many hazardous situations and potential couplings of unfavourable circumstances (domino theory, Figure 2-2). As an example, the electric mains voltage is a hazard. We protect ourselves from it by insulating live parts. A hazardous situation might occur, if the insulation is damaged. However, this does not already unavoidably cause harm. This could only occur due to a further event such as touching this very part. But even this might not cause harm if we were wearing insulating shoes. Damage could occur only in case of additional unfavourable situations. For instance, if at the same time our other hand would touch a grounded part, harm could occur – but even in that case we would still be protected by an additional safety precaution such as the residual current circuit breaker of the electric installation (Chap. 8.5.1). Only if in addition to all other unfavourable conditions just in that very moment the residual current circuit breaker failed could we get injured.

The probability of harm would become only infinitely low if safety precautions would be infinitely good. However, unfortunately, we all know that nothing exists which

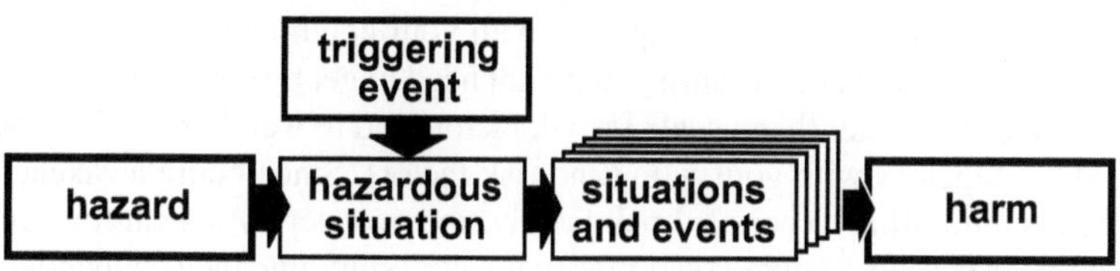

Figure 2-1: Difference between hazard, hazardous situation, and harm

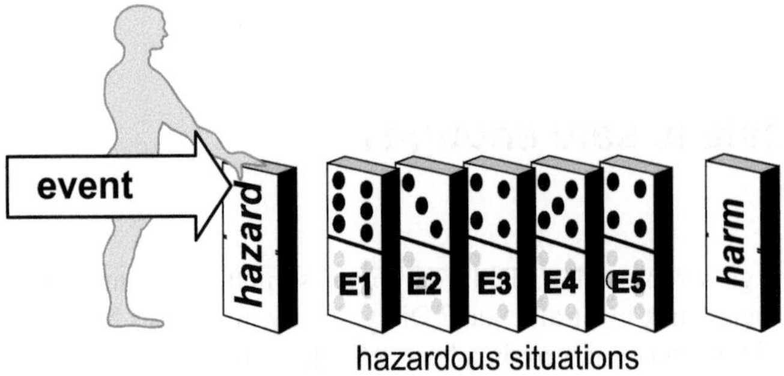

Figure 2-2: Series of events leading to harm

is infinitely good, least of all, for infinite time: no insulation of electric voltage, no sealing of liquids or gases, and no mechanical support. Therefore, we have to recognize that total safety in the initial meaning of the term does not exist. In the real world we have to accept less "safety" and make reasonable compromises.

! In principle, total safety does not exist

The reason why we are protected from daily disasters is that as a reaction to omnipresent risks we have learned to be cautious. In most cases we rely on redundancy in terms of providing a second equivalent alternative in case the first fails. For instance, if it would be very important to get up on time so as to catch the plane to go on vacation most of us would set a second alarm clock (redundant safety precaution) or even in addition ask someone to give a wakeup call (second redundancy) to be on the very safe side.

However, sometimes, it may be like elflock-stricken and things go wrong in spite of all our precautions. This can be explained by an example (adapted from /56/).

An example of daily life

The background scene is unremarkable. Imagine, both of your neighbours have cars, one of them has a dog, a construction site is nearby, a prison is in the town, the bus driver's wage negotiations failed, an interesting thriller is on TV's night program, and your vacation is close, and, therefore, your parents-in-law visited you to discuss caring for your apartment during that time. Your plan for tomorrow is to go for an important job interview and to convincingly present yourself. Therefore, you have taken the following day off. Because of this you plan to sleep longer. Thus, you handed your alarm clock over to your wife asking her to wake you up next morning before leaving.

Since you were still upset from a quarrel with your neighbour because of the enervating barking of his dog and because you do not need to get up early next morning you decided to go to sleep later than usual (1st risk factor) and to watch the TV movie about a jail breakout together with your wife (2nd risk factor). Since being awakened by an alarm clock is unfamiliar to her (3rd risk factor), she overslept. As a cautious man you have asked your parents to give you a morning call (1st redundancy) which woke you late, but not too late. However, your wife had no time to prepare coffee on which you depend in the morning (4th risk factor), and had to borrow your neighbour's car (5th risk factor) to still reach her office in time. Since coffee is important, you decide to have at least instant coffee and take your time to heat some water (6th risk factor). When it is ready you realize that the instant coffee had been used up by your visitors (unexpected coupling of a usually independent event). Because of the time loss and the missing coffee you become nervous (7th risk factor). To make an impression at the interview you decide to change your suit (8th risk factor). While dressing in a hurry, the shoelace tears (a consequential failure of time loss). This further increases your nervousness (a coupling with time loss). After rushing out of your home, the door shuts behind you. At your car you become aware of having left your keys in the suit you wore yesterday (a consequential failure of changing the suit). At first, you are not worried because for such cases there is a spare key hidden below the door mat (2nd redundancy). However, when grasping for it, you remember that this key has been given to your parents-in-law in view of their planned care during your vacation (unexpected coupling to your vacation). Unable to drive your own car, you could have used your neighbour's car (3rd redundancy). However, this had already been lent to your wife (a coupling to oversleeping, in safety terms another consequential failure). On any other day it would have been possible to ask for the other neighbour's car (4th redundancy), but because of yesterday's quarrel with him, this is impossible now (another unexpected coupling). Still optimistic, you decide to take the bus (5th redundancy). However, at the bus stop you are waiting in vain. It turns out that because of the failed wage negotiations public transport is on strike (further unexpected coupling of events). When you try to call a taxi (6th redundancy) you are informed that because of the strike all taxis are occupied (a consequential effect of the strike). As a last resort you try to reach your destination by hitchhiking (7th redundancy), but on that day without success. Later you were told that prisoners had escaped from jail by successfully copying a trick shown on the TV's night thriller (unexpected coupling with an independent event). Therefore, a warning had been issued not to give a lift to any unknown person (a further unexpected coupling with an independent event). When you finally realize that you would be unable to keep your appointment, you decide to at least apologize by phone (8th redundancy). However, unfortunately, you find the line dead. A bulldozer at the nearby construction site had damaged the telephone cable (another coupling with an independent event). This makes your personal disaster unavoidable. Because of the loss of trust in your reliability your chances for this new job are gone and you can forget it.

This example demonstrates that accidents usually don't have one single cause but occur at the end of a series of events. In fact, forgetting the car keys would not have been a problem, if your wife had not overslept, if the quarrels with the neighbour had not happened or if the bus service would have been available. Even all of these things would have been insignificant or would not even have been noticed if just on this day it would not have been so important to be on time.

However, this example also shows that there are always a manifold of independent and initial decoupled situations which in the course of an event may become relevant, influence further development and even contribute to a fatal end. This leads to an important conclusion: Safety precautions can reduce but not fully eliminate risk.

Unexpected couplings make safety technology difficult. Also in medical technology it is impossible to identify or foresee all potential spontaneous couplings and unfavourable situations. Therefore, attempts to provide or enhance safety aim at removing or at least reducing already identified hazardous situations. There remain enough other hazardous situations that we are not aware of. From this follows the basic imperative: not to make compromises but to principally clear hazardous situations once they are identified.

> ! identified hazardous situations must be cleared – there are still enough others remaining

2.1.1 Risk perception

Already the terms risk as such and in particular individual risk perception is complex. In common speech the term "risk" is used in the sense of "probability of occurrence" of an event, such as the risk of getting a parking ticket.

> ! $risk_{common\ speech}$ = probability of occurrence

Besides this, risks are perceived with very individual weighting. Perception can considerably vary from person to person, and even be in contradiction to scientific results. In his play "Lumpazivagabundus," Nestroy, the famous Austrian author lets one of his actors, the shoemaker Knieriem, explain why he does not want to settle down and have a family. He is full of fear because he is deeply convinced that a comet will strike the earth soon and destroy it. Although such an event is extremely improbable, for Knieriem it poses an overwhelming threat and influences his entire life.

In spite of all scientific attempts to quantify risks, our subjective perception of a risk factor is almost immune to scientific data – or do you deliberately travel by train because you know that the associated risk of an accident is significantly lower compared with going by car?

The reason for our distorted view of risk lies in our subjective risk perception which is dependent on a series of individual weighting factors (IWFs)

> ! $risk_{perception}$ = probability of occurrence x IWF1 x IWF2 x IWF3 x ...

By us risk may be dramatically overestimated (e.g. the risk to get a brain tumour from mobile phone use) as well as critically underestimated (e.g. to get lung cancer from smoking or skin cancer from UV tanning). There are several parameters responsible for our distorted view of risks.

For example, when driving a car the risk is perceived as much smaller than it really is. The reasons are, *familiarity* with driving, the existence of *safety means* (e.g. safety belts, airbags), the feeling of *control* through our own efforts (e.g. by adequate behaviour, speed control, quick reaction). In fact, the skill in driving a car seems to be one of the most fairly distributed abilities: One rarely finds anyone claiming to be a bad driver! Further factors reducing perceived risk are the *personal benefit*, *customization* (e.g. to daily reported car accidents), the lower *catastrophic potential* of frequent but few deaths compared to the large number of victims of an airplane crash, and the lower *media attention* of daily incidents compared to the big news of an airplane crash.

It is the opposite with the wide-spread concern about potential adverse health effects of power lines or mobile phone base stations /50/. Although risks could not be proven at exposures below existing exposure limits, and alarming risk estimates are based only on vague hypothetical assumptions, risks are frequently overestimated by the general population. In addition, this attitude is supported by wide-spread *lack of knowledge* of physics, the periodic *high media attention* in combination with sales promotion of protective products, the *lack of trust* in reassuring messages from the authorities, the *inability of perceiving* or *controlling* electromagnetic emissions, the *unclear personal benefit* ("Why power lines? I get my current out of my socket outlet!"), and, finally, the fact that risks associated with technical sources are *man-made* rather than risks from nature which are perceived as unavoidable. Risks are felt to be completely unacceptable once they are *stigmatized,* which is the case in Austria, for example for risks from nuclear power plants (Austria has the only turnkey-ready nuclear power plant worldwide that never became operative).

It is no real contradiction that the same risk factors, for example electromagnetic emissions, are perceived quite differently, namely almost negligibly, if emitted by our own electric appliances or mobile phones, in spite of the fact that they are considerably higher than fields from outdoor sources. The reasons for this are the awareness of evident *personal benefit* and the feeling of *controllability* because one could switch them off at any time. It is not even considered an argument that we do not have any sense allowing us to become aware of such a necessity.

Risk-elevating factors (horribleness, ignorance, inequity and stigmatization) and risk-reducing factors (benefit, controllability, familiarity and risk tolerance) can be summarized in terms of a risk balance as shown in Figure 2-3.

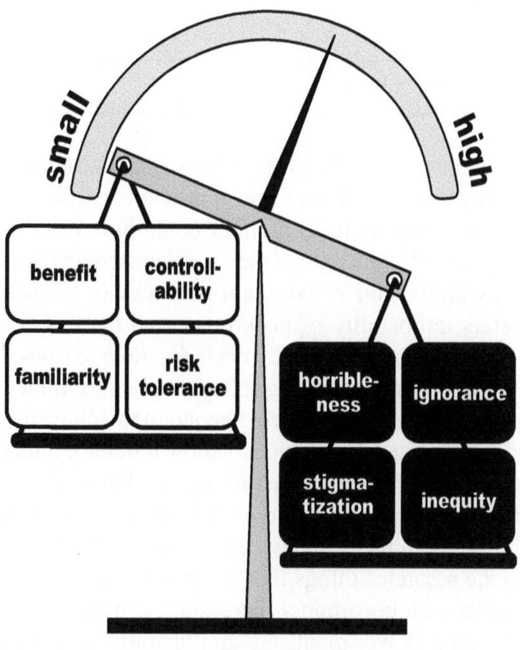

Figure 2-3: Risk balance with factors elevating and reducing perceived risk

Consequences for medical device safety

In regard to medical device safety, the difference between objectively assessed risks and subjective risk perception has three main consequences:

- individuality of risk perception is one of the reasons why risk analysis and assessment should not be performed by one single person but by a team of individuals with (usually) differing risk perception;
- individuality of risk perception is one of the reasons why standards reduced the number of concrete safety requirements. For many risks identified by risk analysis it is left to the manufacturer to decide whether from his point of view they are acceptable or not;
- the safety standard EN 60601-1 accepts that results of risk analysis and assessment may differ depending on moral and ethical concepts of a society and/or cultural area /27/. However, this new freedom may be counterproductive and undermine global harmonization because risk perception and assessment differ and risks accepted in one region of the world might be considered unacceptable in other regions: Therefore, devices that are considered acceptable in an area with high risk tolerance could encounter future trade barriers when introduced into regions with lower risk tolerance.

2.1.2 Objective risk

In technology risk is clearly defined. It accounts for the probability of occurrence of an adverse effect and the caused harm. Therefore, "risk" is understood as the product of probability and harm /21/.

! $risk_{objective}$ = probability x harm

As an example, a breakdown of an ECG recorder would cause only inconvenience to the patient, as the patient would have to return at another time for the investigation, while the same event involving an ECG monitor could have even lethal consequences if for example cardiac asystoly in a patient was not detected and the alarm not given.

To determine risk with the above formula it is necessary to quantitatively estimate two relevant parameters, probability of occurrence and harm. However, this is more difficult than it might appear at first glance. This is for three reasons. First, that rareness of adverse events increases with increasing safety and due to limited cases so does the uncertainty on the probability of occurrence. Second, new innovative products by definition are associated with limited experience to allow reliable estimation of risk parameters. Third, difficulties in completely identifying, analysing and assessing potential failures increase with increasing complexity of failure possibilities of interacting parts, components and modules composing a device (see Sect. 2.2.1.2).

Therefore, it's in the nature of things that "objective" risk assessment cannot result in more than an estimate with uncertainty increasing with decreasing frequency of occurrence of events. Nuclear power plants are an illustrative example. Initially, the risk of a "maximum credible accident" associated with a worst-case scenario was estimated.

Years later, experience showed that even worse scenarios were possible. Consequently, estimates were made for a "super maximum credible accident" which demonstrates an interesting fact for linguists, namely that bad/worse/worst can be even further enhanced to the term "super-worst."

2.2 Risk management process

Among others the third edition of the generic medical devices safety standard EN 60601-1 /27/ differs from previous versions by replacing specific technical requirements by sole descriptions of protection goals. Now it is up to the manufacturer to analyze and assess risks by considering the individual characteristics of his device, and to take care of measures considered necessary to control them. This must be done by planning, implementing and maintaining a structured and documented risk management process according to EN ISO 14971 /21/.

Death trap hospital bed

Bonn: Experts warn about underestimated hazards. Within one year, several deaths occurred in hospitals, nursing homes and homes caused by medical beds. Reasons were electric shocks and burns due to damaged cables, smoke poisoning due to smouldering fires, contusions, strangulations and falls.

It is no longer sufficient to design and produce medical devices to the best of one's knowledge. Now it is also required to systematically analyze, assess, control, and monitor risks to ensure that sufficient protection is provided against ***all reasonably foreseeable hazards***.

First, this leads to the question what should be understood by "reasonably foreseeable" and how it could be assured that all such hazards are identified and their risks efficiently controlled.

It is remarkable that accidents are not mainly caused by high-risk devices. Devices with inherent hazards originating from their method, performance or application are usually thoroughly analyzed and carefully applied. However, awareness of potential hazards of apparent low-risk devices is frequently low and their risks underestimated. This is the only explanation why each year patients die or are severely injured by hospital beds or why simple bedside lamps become lethal time bombs.

A "process" is much more than just a "procedure" that consists of a set of defined instructions. It comprises a network of interacting activities and feedback loops. The difference in these terms resembles the difference between a controller (procedure) and a multi-loop control system (process). The risk management process includes activities spread over the whole product life cycle, from the choice of the applied method, development of a concept until design, realization, production, marketing and market surveillance (observation of application). It involves cycles of redesign and product type improvement until the end of production (Figure 2-4). The life time of an individual device starts with manufacturing and involves marketing, application and maintenance, and extends until its disposal at the end of its service life.

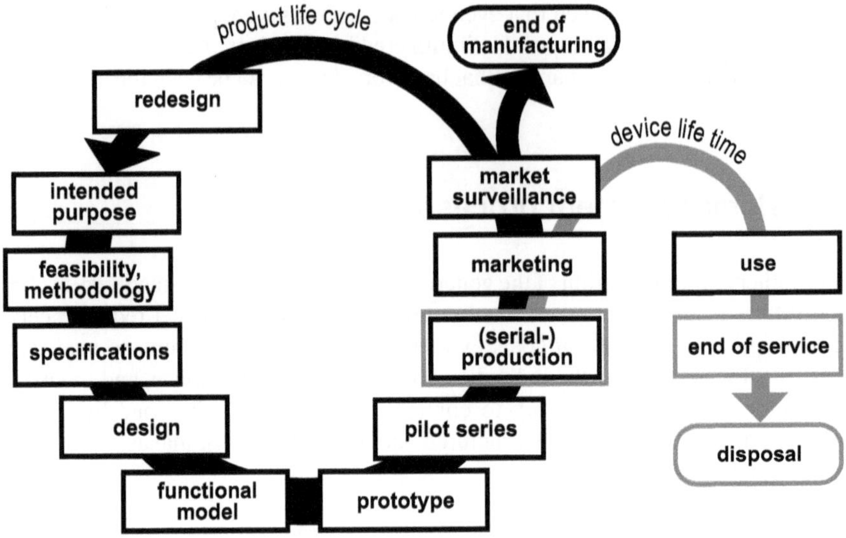

Figure 2-4: Product life cycle and device life time

Manufacturers must plan, implement and maintain the risk management process. In this process risks must be identified and assessed that may arise from intended use as well as from single fault conditions and foreseeable misuse and error. Decisions must be taken to reduce and control risks. Afterwards it must be tested whether decided safety precautions were realized (verification) and then, it has to be assessed whether they were sufficiently efficient to assure the device meets the essential requirements (validation) and exhibits the required low risk level (Figure 2-5).

The risk management process is much more than just the risk analysis that was required from manufacturers so far. A schematic summary of these activities is shown in Figure 2-6. It comprises

- ***organizing*** risk management in terms of
  - identifying safety goals, criteria for acceptance or rejection of single risks and the total risk, and criteria for initiating correcting actions,
  - defining responsibilities and authorizations and
  - providing sufficient resources, personal and financial.
- ***planning*** risk management by elaborating a plan which considers complexity, methodology and timing of risk analysis, risk assessment, risk reduction and risk control including verification and validation of risk control measures. Moreover, post-manufacturing activities have to be specified including monitoring use and performing market surveillance.
- ***risk analysis*** which systematically identifies reasonably foreseeable hazards and associated risks under normal condition and single fault condition, during intended use and foreseeable misuse and error.
- ***assessing*** identified risks.
- ***controlling*** risks, analyzing options for action, deciding on correcting measures, and analyzing their potential adverse retroactions.

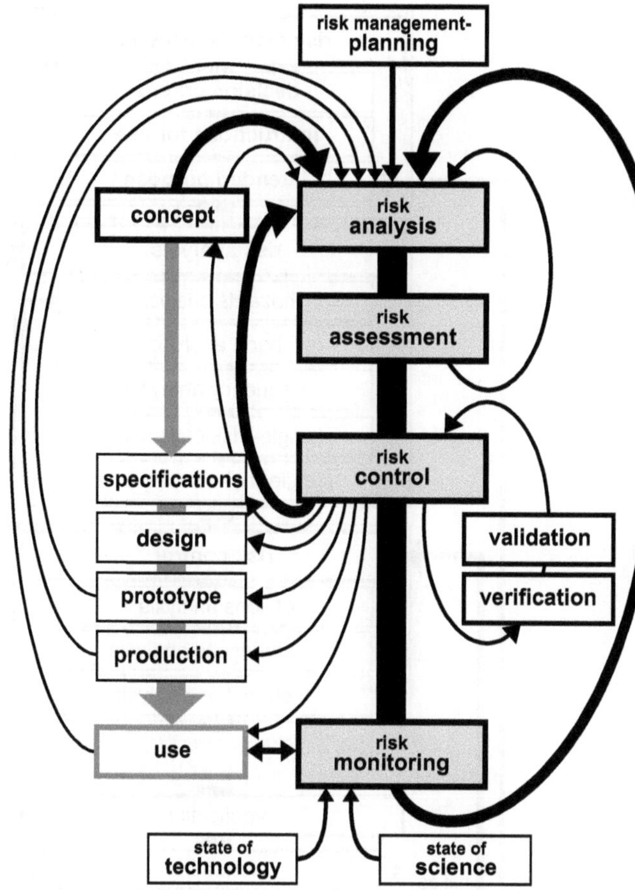

Figure 2-5: Risk management process with the key elements risk analysis, assessment, control and monitoring

- assessing ***residual risks*** based on defined criteria for acceptance and rejection.
- ***verifying*** realization of intended risk reduction measures.
- ***validating*** the efficiency of intended risk reduction measures.
- ***assessing*** the overall risk, if necessary performing a risk/benefit analysis.
- ***controlling*** and ***monitoring*** risks by actively acquiring and analyzing internal and external data associated with experience of use, and analyzing and assessing this information in regard to its impact on risk analysis and risk control.

2.2.1 Risk analysis

Systematically performing and documenting risk analysis should already have become a self-evident element of product development, similar to generating a circuit diagram or a layout of an electronic board. Now, risk analysis accompanies the whole product life cycle (Figure 2-7). It begins with the product idea to assess feasibility, at which point it may already lead to consequences in terms of product design, limiting the in-

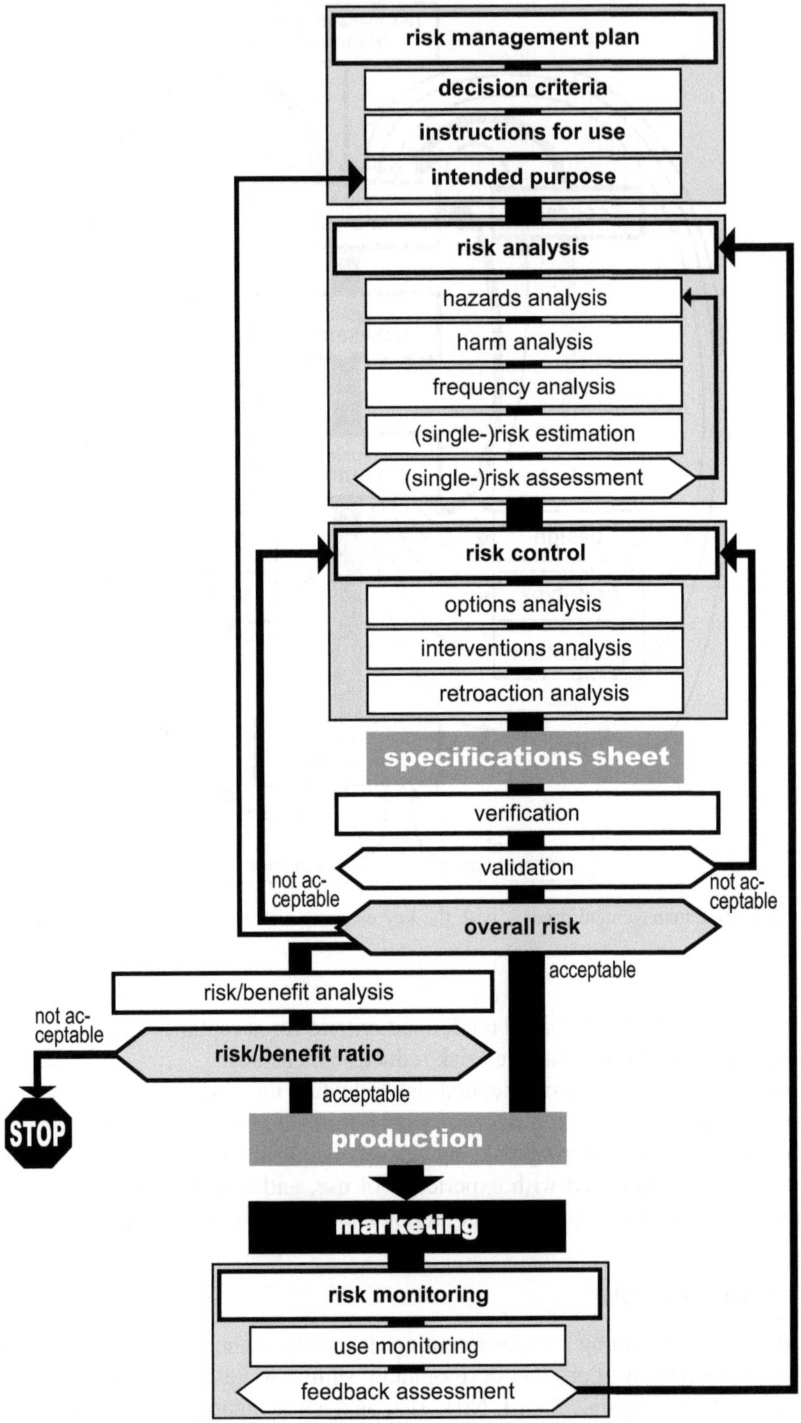

Figure 2-6: Elements of a risk management process

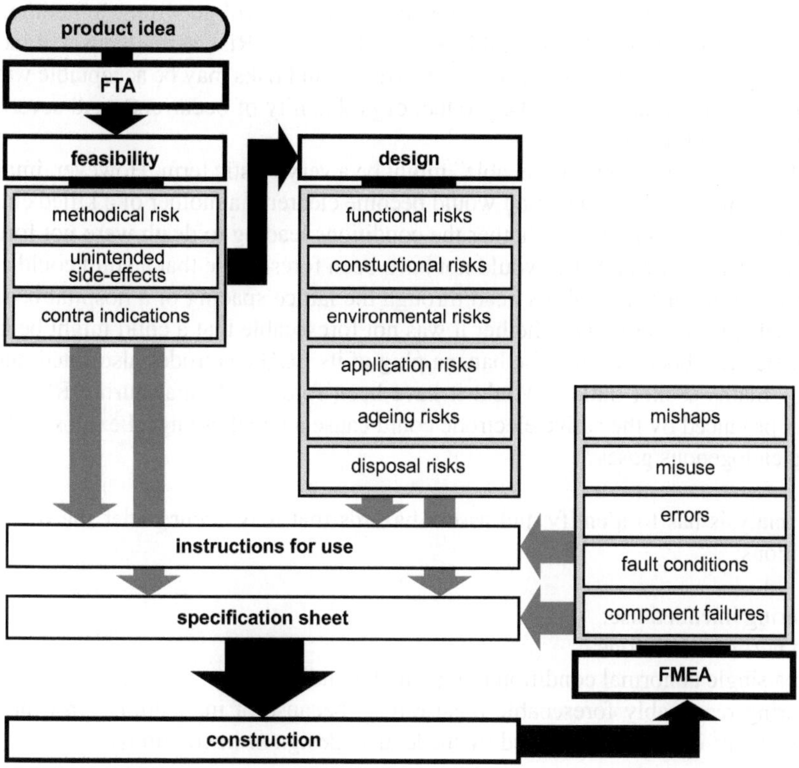

Figure 2-7: Risk analysis during device development using fault tree analysis (FTA) and failure mode and effect analysis (FMEA)

tended purpose, use conditions and the target group of patients. In the successive design phase, risk analysis influences the specification sheet. After product realization failure mode and effect analysis (FMEA) of hard- and software components is performed and potential impact of manufacturing processes analyzed. Finally, risks associated with distribution and storage are identified, and incidences registered and analyzed which may be encountered during use. Market surveillance data are analyzed to check whether initial estimations on probability of occurrence and severity of harm are still valid or have to be adapted.

The first step of risk analysis is identifying ***all*** reasonably foreseeable hazards which may arise during normal condition and single fault condition including human error, misuse and mishaps. This is a very ambitious goal and should not be underestimated. Risk analysis is not a one-man-job which can easily and quickly be done, but should be an intensive brainstorming process of a team. Such an interactive process of individuals benefits from complementary contributions which are triggered by individual's different risk perceptions. Therefore, it is essential not to stop brainstorming too soon, and not before the pool of associations and ideas is exhausted.

Brain "storming" means allowing free play of contributions and inspirations without premature self-censorship. Therefore, none of the spontaneously mentioned ideas about potential hazards should be ignored even if it might appear curious. Murphy's law should be taken seriously: everything that can go wrong will go wrong sometime, everything that can be thought of, could happen. Be patient. Risk assessment will include weighting harm with its probability of occurrence, and risks may be acceptable without further precaution, anyway, if the product of probability of occurrence and severity of harm is sufficiently low.

The term "reasonably foreseeable" might be a very elastic term. However, imagine, after a lethal accident its meaning would become clearer if a mother of a killed child or the judge in a lawsuit asked whether the conditions leading to death were not foreseeable. For instance, whether it would not have been foreseeable that a child could strangle itself after slipping with its head through the lattice spacing of a hospital bed, and was unable to retract it? Or whether it was not foreseeable that a child might be killed by an electric shock because the banana plug of its ECG electrodes also fitted into the nearby mains socket outlet? Would it have been foreseeable that during RF surgery sparks produced by the active electrode could cause a lethal laryngeal explosion by igniting endogenous gases?

Risk analysis has to identify and assess hazards that may occur under the following conditions:

- during intended use;
- in normal conditions;
- in a single abnormal condition (single fault condition)
- during reasonably foreseeable misuse (e.g. because of inattention, error, neglect, ignorance or misuse provoked by inadequate design or instructions)

Remark: The difference between hazard, risk and harm is discussed in Sect. 2.1.

In particular, the following sources of danger should be analyzed:

- ***method*** (e.g. galvanic connection of the patient to an electric circuit, delivery of critical substances, energy, radiation, application of pressure or heat);
- ***function*** (e.g. life-saving, life-supporting, life-sustaining, monitoring vital parameters, measuring, controlling, treating, emergency use);
- ***construction*** (e.g. component failure, leakages, combination of risk factors such as electricity, oxygen, flammable gases, material ageing, abrasion, dependency on external electric installation, supply with cooling agents, driving pressure etc.);
- ***patient*** (e.g. age, general condition, ability to react, move, perceive, device-dependency, dementia, contraindications);
- ***unintended side-effects*** (e.g. nerve or muscle stimulation, electroshocks, tissue burns, vessel ruptures, poisoning, contusions, breaks, fires, explosions, electromagnetic interference);
- ***accessory*** (e.g. relevance for safety, accuracy and reliability, its durability, availability, suitability for disinfection and sterilization);
- ***environment*** (e.g. impact ***by*** environment: humidity, temperature, pressure, sunlight, mechanical load, electromagnetic interference;
  impact ***on*** environment: leakage, diffusion or emission of hazardous substances during use or after disposal);

- ***user*** (e. g. critical application, knowledge, stress, attention, diligence, error, stress, ignorance, misuse);
- ***operator*** (e. g. installation, service, maintenance, recurrent testing).

To reliably identify all relevant aspects hazard analysis should be performed as systematically as possible. The sequence of events should be followed until all possible causes are identified. Option analysis for risk reduction should consider their different efficiencies and potential retroactions.

There are two main approaches to risk analysis (Figure 2-8):

- ***fault tree analysis*** (FTA): It starts with the harmful event and goes back step-by-step to all potential initial causes (top-down approach);
- ***failure mode and effect analysis*** (FMEA): It starts with the potential failures and defects and follows their consequences step-by-step along the causal cascade until the final harmful event (bottom-up approach).

2.2.1.1 Fault tree analysis

At the beginning of a product life cycle when the feasibility of a product idea needs to be checked, the question arises which hazards may be associated with the new product and whether they could be managed. For this case the fault tree analysis is recommended. Table 2-1 shows an example of the protocol of a systematic FTA. The protocol should be filled out from left to right. Starting with a list of sources of danger, the associated potential harmful events are identified for each of the listed items. An analysis is then made of what needs to happen to cause harm to patient, user and/or environment by asking the question "what must happen … to cause this harm?"

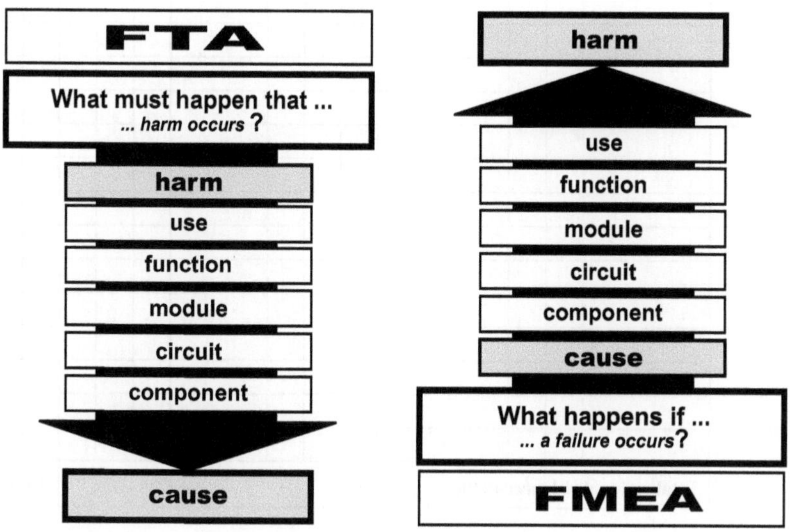

Figure 2-8: Risk analysis approaches: Top-down by fault tree analysis (FTA), left, and bottom-up by failure mode and effect analysis (FMEA), right.

> **!** What must happen ... to cause this harm

Then the probability of occurrence is estimated and the resulting risk determined. Next, potential precautionary measures to avoid or reduce the risk are discussed and their efficiency assessed. Finally, the risk remaining after implementing risk precaution measures is estimated and potential unintended retroaction of the chosen measures assessed.

Table 2-1: Example of a risk analysis protocol

	Risk Analysis	**Version:**
Product		**Date:**
Intended use		**Examiner:**
		Approved:

Source of danger	**Hazard**	**NC**	**SFC**	**Harm**	**P**	**H**	**R**	**Precaution**	**FB**	**SL**	**A**	**P**	**H**	**RR**
Method														
Function														
Construction														
Patient														
Side-effects														
Accessories														
Environment														
Ecology														
User														
Operator														
	Total residual risk acceptable:				**yes O**				**no O**			**TRR:**		

NC ... normal condition, SFC ... single fault condition, R ... risk level, P ... probability of occurrence, SL ... safety level (1,2,3), A ... accepted measure, FB ... feedback, RR ... residual risk level, TRR ... total residual risk level

Example: Nerve and muscle stimulator, output current up to 80 mA:

- *source of danger:* electric current directly flowing across the patient;
- this occurs during normal condition and single fault condition,
- the output current 80 mA is associated with the hazard of heart fibrillation,
- the *harm* could be death of the patient;
- *probability* of occurrence is determined by the probability of the electric pathway across the heart and the hazardous output setting. Considering that no specific safety precaution is already foreseen, the probability could be estimated with "sometimes" or even "frequent" (Chap. 8.1.3).
- with the aid of the risk matrix (Figure 2-9) fibrillation risk can be assessed to be unacceptably high (risk level 1).

The decision, whether a particular risk-reduction means should be realized and which of the various options should be chosen to reduce a risk, is up to the manufacturer. However, he needs to be aware that different options may have different efficiency (safety level) and that he is legally obliged to stick to the given hierarchy and to prefer constructive measures (inherently safe design) over protection means (conditional safety) and warnings (see Sect. 2.3.3).

In the example of the nerve and muscle stimulator the following risk reducing options could be considered:

- *direct safety* (safety level 1) by constructively limiting the output current to an inherently safe level (e. g. 10 mA) which technically excludes heart fibrillation;
- *indirect safety* (safety level 2) by activating an alarm in case safe output current values are exceeded, thus enhancing attendance and risk awareness;
- *indicative safety* (safety level 3) by affixing warnings to the device and in the instructions of use that raise awareness of heart fibrillation risk in the case of high output currents and/or inappropriate placement of electrodes.

Means of protection also need to be analyzed as to whether they may themselves create new risks due to unintended retroactions. As an example, cutting tissue with high electric current densities would be easily possible with 50 Hz mains currents. However, unacceptably high risk would follow from the side-effect of unintentionally stimulating muscles which would lead to uncontrolled convulsions of the unconscious patient. An efficient remedy preventing such effects would be increasing the electric current frequency above the cellular stimulation limit of about 100 kHz (see Chap. 8.1.3). In fact, for this reason surgeons apply RF surgery instead of mains frequency surgery. However, this solution leads to an unintended retroaction. At such high frequencies RF electromagnetic fields are emitted that could interfere with other electromedical devices and even might lead to health-relevant overexposures of medical staff. Therefore, additional precautionary measures need to be taken to minimize risk from this unintended retroaction of the precautionary measure.

2.2.1.2 Failure mode and effect analysis

Once the device has already been designed and/or a functional sample or prototype realized, risk analysis aims at systematically identifying hazards originating from hardware and software elements in normal and single fault condition. For this purpose the failure mode and effect analysis has proven useful. Starting from the various basic elements an investigation is carried out to identify which failures with which consequences could occur. Each (safety-relevant) component is checked by asking the question "what happens, if a failure occurred?

> **!** What happens, if a failure occured

To identify hazards it is necessary to follow the causal cascade step-by-step from the basic component to the circuit, module and function to the final harmful event. However, in particular in case of very complex devices such an approach would cause an unacceptable effort. For this reason the combined application of both approaches is chosen.

2.2.1.3 Combined failure analysis

To limit the effort for FMEA to a reasonable amount, a two-step approach is useful: In the first step FBA is used to identify safety-relevant modules, circuits and components. In the second step FMEA is performed but now restricted to such identified safety-relevant elements only.

2.2.2 Risk assessment

2.2.2.1 Single risks

Risk levels

In general, only in exceptional cases can risk of medical devices be quantitatively determined. However, assessment can be done qualitatively by assigning the two risk parameters, namely probability of occurrence and severity of harm to verbally characterized categories. As an example, probability of occurrence can be assigned to the categories "frequent/sometimes/occasional/seldom/unlikely/unbelievable." However, an event can never be totally excluded even if classified as having unbelievable low probability of occurrence. The reason is Murphy's Law of experience which states that everything that ***can*** go wrong ***will*** go wrong – sometime.

> **!** Everything that can go wrong will go wrong

Therefore, if manufacturers refuse to correct deficiencies with the argument "so far, nothing ever happened" it must be stressed that this is, of course, no evidence that the associated risks would not exist at all. Even if the company had a total feedback system and unreported cases could be excluded (which is a difficult to assure) this deceiving argument just means that probability of occurrence of harm is lower than the value derived from the available observation time of the devices already put on the market – and this might not make much impression in the case of newly developed devices or devices sold in small numbers and/or seldom used.

Severity of harm, although difficult to express in numbers can fairly well be assigned to verbal categories such as „small/medium/severe/catastrophic."

The limited numbers of categories of both risk parameters lead to a limited number of their possible combinations (risks). This allows creation of a risk matrix (Figure 2-9). Consequently, this bulk of risks can be classified in regard to risk acceptability which results in risk levels, ranging from unacceptably high risks (risk level 1), undesirable high risks (level 2) and justifiable risks (level 3) to tolerable risks (level 4).

Risk assessment

Because of the diversity of medical devices and their different benefits it is not possible to develop a universal rule for accepting risk, although some guidance can be given (Sect. 2.2.3). However, there is general agreement that attempts to reduce risk must be intensified with increasing frequency and/or severity of harm. On the basis of this principle the following risk levels can be defined (Figure 2-9):

Risk level 1: This level comprises unacceptably high risks. Such high risk could only be justified in exceptional cases if less risky alternatives are not available and risk/benefit analysis demonstrates a sufficiently high benefit.

Risk level 2: This level comprises risks that are acceptable only if the benefit is sufficiently high and all attempts have been made to minimize residual risks to an extent

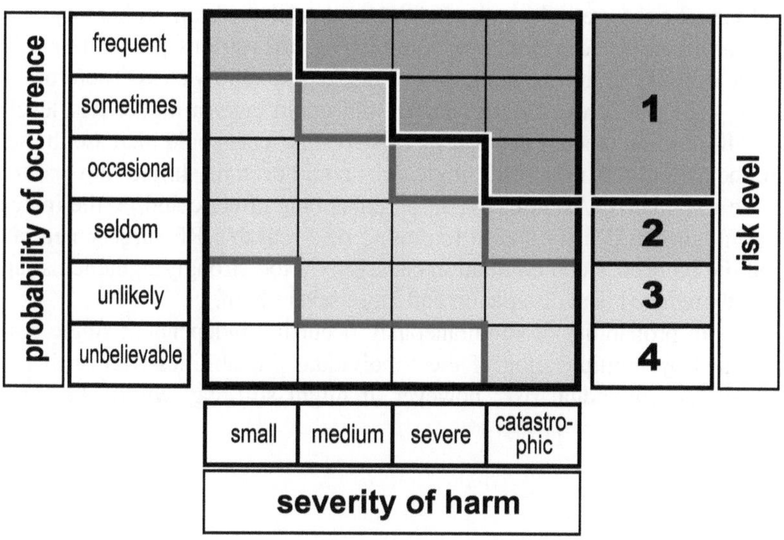

Figure 2-9: Risk matrix formed by probability of occurrence and severity of harm, with risk levels 1 to 4

that could be achieved with reasonable effort. This approach is known as the ALARA principle (as low as reasonably achievable).

Risk level 3: This level comprises risk that could be accepted if all attempts have been made to reduce risks with economically justifiable means. This approach is known as the ALARP principle (as low as reasonably practicable).

Risk level 4: This level comprises risks that are low enough to be generally acceptable as a single risk. However, as a general principle and in view of the (accumulated) overall risk further reduction should be aimed at if this is possible with simple and cheap means.

2.2.2.2 Overall risk assessment

Risk analysis leads to a number of individually assessed single risks. Even if many single risks might be acceptable on their own it does not necessarily mean that their cumulation will be acceptable. It is like a single bee-sting which might be unpleasant but could be tolerated, while many of them could even be lethal if they occur in a short time. Likewise, the simultaneous presence of many risks, although acceptable in each single case, could lead to an unacceptable overall risk. Therefore, overall risk assessment requires analysis, to determine how many of the single risks are associated with independent events, whether single risks could enhance each other, and how their increased number affects the probability of occurrence of such a risk. Like the chance of winning increases with the number of lottery tickets, the probability of occurrence of harm increases also with the number of individual risks. Therefore, it has to be considered whether the multiplicity of single risks changes the assignment to a probability class or whether the interaction of single risks could increase the severity of harm. It must be kept in mind that the assessment period extends to the whole expected service lifetime of a device.

If individual risks are independent of each other, the overall probability of occurrence of harm can be determined by summing up the various probabilities. Consequently, the overall risk may be assigned to a higher probability class. Therefore, it may stay in the same column of the risk matrix but could enter a higher risk level (Figure 2-10). If the simultaneous occurrence of adverse events could increase the severity of harm, the overall risk of several single risks could even be shifted into another column of the risk matrix. As an example, an enhancing effect could be the increase of intracorporal current density due to loosening of a reusable RF surgery neutral electrode and the reduced blood circulation caused by a too strongly tightened strap with consequently reduced heat dissipation and more severe burn.

The overall probability of simultaneously occurring independent single risks can be determined by multiplication of their individual probabilities. Hence, it is lower than that of any individual risk; however, it might still stay within the same risk class.

Overall risk matrix

The overall risk depends on the number and interrelation of individual risks, the kind of device and the duration and frequency of its use. Increasing numbers of individual

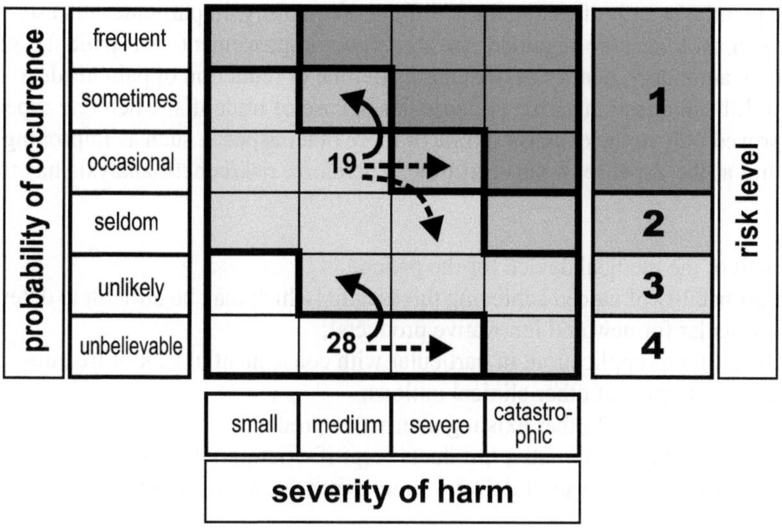

Figure 2-10: Risk matrix for overall risk assessment considering multiplicity of single risks and their potential interaction leading to potential changes of probability and/or severity of harm

risks may result in the necessity to further reduce initially tolerated single risks. Since it is not possible to derive a general rule, assessment has to be made case by case. However, it is recommended to generate a risk matrix for overall risk assessment, where as a first step the sums of identified clusters of individual risks are entered in associated matrix elements (Figure 2-10). In a next step it is investigated whether it is necessary to shift the overall risk into a higher class of probability and/or a higher class of severity of harm. The criterion, how many individual risks should be considered necessary for changing risk classes should become stricter with increasing severity of harm and frequency of occurrence. This approach is symbolically demonstrated in Figure 2-10.

2.2.3 Risk/benefit assessment

Whether the overall risk of a device may be acceptable is not only dependent on the benefit but also on the available alternatives. If solutions already exist, which provide a similar benefit with much lower risk, the decision to accept a higher overall risk could be negative. Assessment of risk/benefit ratios is of particular importance, if the overall risk still remains high in spite of all attempts at further reducing and minimizing it. This is particularly important if on its own an overall risk might be considered unacceptably high. In general, this would require stopping device development. However, in exceptional cases, it might be nevertheless justified to realize the product, if other alternatives were missing or would have even higher risks. For instance, initially it was justified to provide heart patients with an artificial heart, to bridge the time gap to transplantation in spite of the life-threatening risk of an infarct because the alternative of non-application would have led to death with higher probability in even shorter time.

Frequently, benefit of treatment is difficult to quantify, in particular in cases where health can no longer be regained. In such cases improvement of a patient's general condition, autonomy, quality of life and/or lifetime or reduction of pain needs to be appraised. Ethical questions arise in particular in case of tradeoffs, when one aspect can be improved only at the expense of one or more other aspects, such as improving quality of life at the expense of survival time. Therefore, risk/benefit analysis has to consider

- benefit of the medical device for the patient;
- the probability of indeed achieving this benefit (which may be difficult to determine, in particular for new and innovative products);
- the risk of non-application, in particular with consequential lack of treatment;
- risk/benefit ratios of other clinical options;
- risk/benefit ratios of other existing alternative products;
- the availability of alternative products (e. g. if existing alternative products might not be available because of delivery time, expenses, applicability etc.).

However, it is the nature of risk/benefit analyses that they don't hold for ever. Progress in scientific knowledge and technical feasibility may make new solutions available that challenge existing devices and make it necessary to update their risk analysis (Figure 2-11).

The difficulty in quantifying risks and benefits can be overcome by again choosing a qualitative approach and differentiating verbally described categories. Benefit could be graded qualitatively into "life-saving/high/considerable/moderate/small/negligible" while overall risk could be classified as "life-threatening/high/moderate/small/

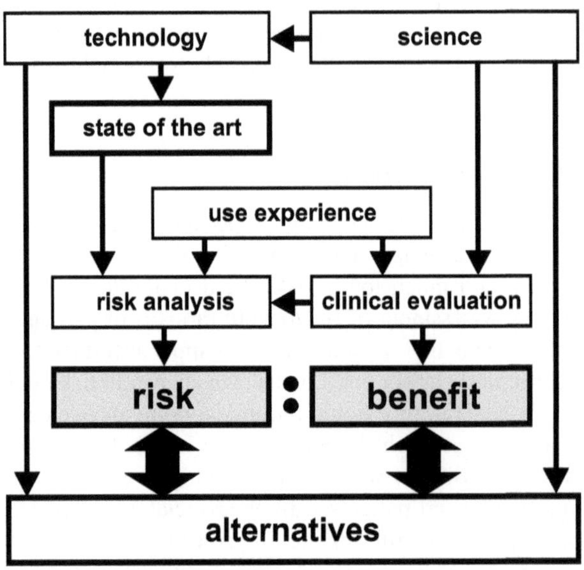

Figure 2-11: Factors influencing risk/benefit assessment

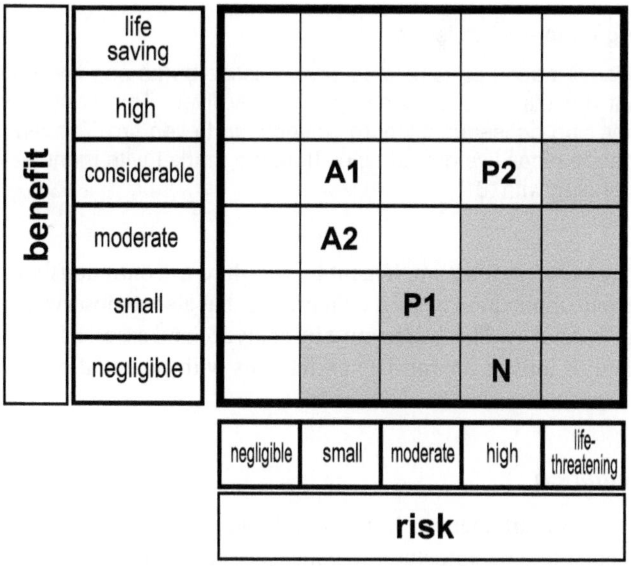

Figure 2-12: Risk/benefit matrix. P1, P2 … product version 1 and 2, A1, A2 … already available alternatives 1 and 2, N … non-treatment

negligible." This again allows creating a risk/benefit matrix where results of the overall risk analysis and assessment results of other alternatives can be entered (Figure 2-12).

An example is shown in Figure 2-12. It is assumed that non-treatment (N) would be associated with high risk, and benefiting from potential self-healing negligible. In comparison with this situation, the product version P1 would offer a small benefit at moderate risk and would be acceptable if there were no other choices. However, in view of existing alternatives A1 and A2 the risk/benefit ratio might not be good enough. The high risk of product version P2 could be justified by its considerable benefit. However, in regard to alternative A1 which offers the same benefit at a small risk the risk/benefit ratio of P2 would probably not be acceptable. If the option A1 would not exist, depending on the risk perception of the manufacturer the high risk of version P2 could be accepted because of its higher benefit compared to alternative A2 which offers only moderate benefit, although to a much smaller risk compared to P2.

2.2.4 Risk monitoring

The risk of a medical device is not only dependent on design, construction and failure probability of components. It is also influenced by the manufacturing process, distribution, use and maintenance. Since risk analysis must be mainly based on estimates, an essential element of a risk management process is to monitor whether these estimates hold in practice. During actual use it could turn out that harm could occur more frequently or be more severe than initially assumed. There is also the possibility that additional risks may be encountered that initially were not identified.

Heating pads – a hot issue

Michigan: Heating pads caused severe tissue burns and fire. It turned out that the supply cable was not properly fixed. This caused short-circuiting and consequently burn wounds, and even fire. The manufacturer had to recall five device types from the market with manufacturing codes ending with "01".

Therefore, manufacturers must implement a procedure to continuously collect data on manufacturing and application of their own product but also to observe experience with other comparable devices. These data must be analyzed and assessed as a control of the risk analysis and to initiate appropriate corrections with potential product redesign if necessary.

Market surveillance

Today, it is no longer sufficient to just passively wait for vigilance reports. There are many avenues open to a manufacturer to gain access to internal and external data. The required effort depends on the risk (conformity class) of a medical device and the experience already on hand concerning its use. However, it is not sufficient to restrict observation to construction and manufacturing only. Even severe danger can arise also for subtle reasons such as erroneous packaging, wrong labelling or misleading instructions for use.

Fatal packaging

Massachusetts: Syringes filled with 100 ml insulin were erroneously packed into packages intended for 40 ml syringes. Since overdosage may lead to adverse health effects including death the manufacturer had to call-back the related batch.

As an example, syringes filled with insulin had to be called back from market because of the risk of dangerous overdosage. The reason was, that syringes containing 100 ml insulin had been put into packages labelled with 40 ml. Market recall was also necessary for non-sterile radio frequency denervation electrodes which erroneously had been marked as sterile.

For market surveillance manufacturers can make use of several sources of data, internal and external (Figure 2-13). A first cheap and easy approach is to assess internal data, already available to a manufacturer – provided there are procedures in place to actually collect, analyze and assess this data for risk monitoring.

- ***internal*** data are unavoidably generated in the course of business. Such data should be analyzed as a matter of course, independent of the conformity class or the inherent risk of a device. They are not only useful for risk monitoring. They also allow detection of degradation of the product, deficiencies of the manufacturing process or unreliable suppliers, for example by analyzing trend curves of various parameters. Useful internal parameters may be

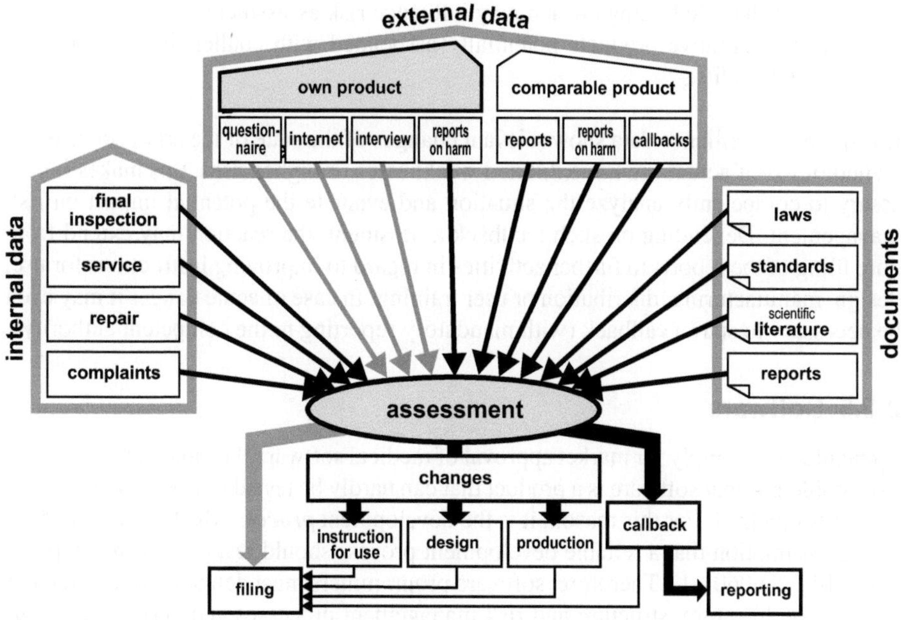

Figure 2-13: Risk monitoring through post-manufacturing activities

- the reject rate of final inspection including their reason. This allows concluding on weaknesses of the manufacturing process and/or of components, for example from external suppliers;
- the number of services, repair, complaints and liability cases;
- deficiencies encountered during recurrent safety tests and periodic inspections of devices already in use. This allows concluding on risks from design, manufacturing, application and maintenance;
- the number of vigilance reports including mandatory reports on severe incidents. They could help in the discovery of yet unknown hazards.

• ***external*** data such as on experience of use usually are only accessible with increased effort and costs. Expensive approaches may be justified if a device is novel and/or has increased risks. There are different possibilities to acquire external data:
  - by motivating clients, customers and/or users to give feedback, e. g. via the internet or service hotlines;
  - by active inquiry of distributers and/or users about their experience with the device (e. g. by questionnaire, personal interviews or telephone interviews);
  - by looking for reports on experience of use, encountered risks, incidents or vigilance cases of comparable devices;
  - by looking for reports on callbacks of comparable devices, for example on homepages or via health authorities.

  In addition, due diligence requires also continuously monitoring other external data such as
  - new standards and regulations defining a (new) state of the art. They may require adaptation of the risk analysis and device redesign;

- new scientific findings which may influence risk assessment;
- new alternative products potentially associated with challenging low risks or high benefits.

If acquired surveillance data show relevant changes of the state of the art of science and technology or if actual use has exhibited weaknesses or higher risks, this makes it necessary to competently analyze the situation and evaluate the potential impact on risk management. Depending on such feedback assessment, the reaction may extend from pure filing of the reports to further activities in regard to improving instructions for use, design, manufacturing, distribution or user training. In case of acute danger it may even be necessary to start a callback (with mandatory reporting to the competent authority).

2.2.5 Software

If manufacturers apply for market approval of medical software they may fall into a trap. The problem is that software is a product that can hardly be tested with reasonable effort once it is finalized. For this reason, it is the development ***process*** which is tested, based on the assumption that a reliable development process should result in a reliable product (§ 14 EN 60601-1). Therefore, software programmers must define in timely manner software architecture, structure and risk management processes and perform continuous documentation. To start this only after finalization on request of a notified body would not only require considerable effort, it would also not follow the basic strategy for product compliance assessment. Therefore, software development must be governed by the risk management process of EN ISO 14971 /21/ and EN IEC 62304 /34/.

In comparison to hardware products software can easily be changed, adapted and/or further developed. This requires continuous updating of records, implementing structured processes, managing of documents and software versions including indentifying, marking, testing, approving and releasing them.

The software development process including risk management must be planned and maintained involving all stations in the software life cycle (see the “V”-diagram in Figure 2-14). First, software architecture is developed and visualized by a flow chart. Tasks are split step-by-step into subtasks until finally small modules are generated which can be directly checked and verified. Afterwards verified components are integrated step-by step into the complex system which is verified and validated and finally (if necessary after certification by a notified body) put on the market.

Reliable functioning of software usually depends on interaction with external software components such as operating system software, software for evaluation, display and communication. It may also process data generated from other software. Such interactions must be included in risk analysis and risk management. Software-specific aspects of risk management are

- potential impact of interactions of computer and data networks, of operating system software and auxiliary software under normal and single fault condition;
- impact of electromagnetic interference on transfer of data and commands;
- erroneous data or data formats, or missing data;
- loss of data files;
- unexpected interference by third-party software;

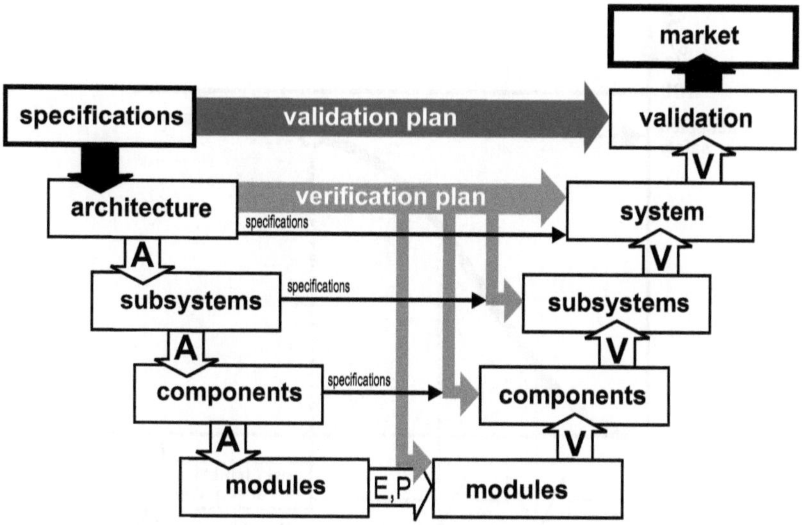

Figure 2-14: "V"-diagram of the software development life cycle. Tasks are split into subtasks (A) until they finally become small modules which after checking are verified (V) and integrated step-by-step into the complex system which after validation is finally put on the market

- attack by viruses;
- impact of changes of external software such as due to automatically generated updates;
- unintended side-effects of own updates;
- random interference with internal or external influences;
- software bugs;
- wrongly timed sequences;
- data safety;
- data protection.

2.3 Medical devices safety

It has already been shown that it is impossible to achieve total safety, but we know that with increasing effort safety can be increased such as by doubling insulation, enforcing pipe gaskets, implementing watchdog routines or intensifying safety checks. However, safety is not available free of charge. It must be paid for in different ways

- by increased ***costs*** (e.g. for cars with seat belts, airbags, antilock brake systems, electronic break force distribution, electronic stability control, anti-collision systems, driver assist systems etc.);
- increased ***discomfort*** (e.g. breathing mask, radiation protection skirts, protective goggles, noise protection);
- increased ***time*** (e.g. safety checklists, watchdog routines, redundancy).

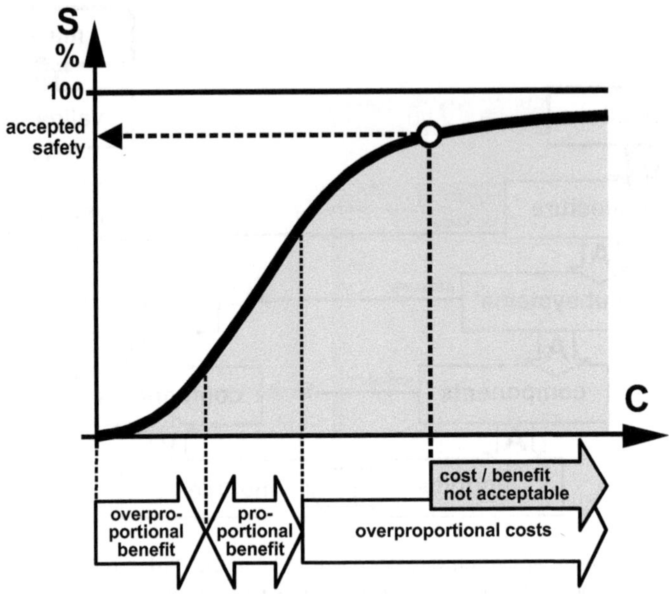

Figure 2-15: Cost (C)–safety (S) curve

We all know that safety needs to be paid for, but we are not willing to spend unlimited money – do you buy only that car that offers all possible technical safety measures irrespective of its price – or do you accept a compromise?

! Safety must be paid for

What applies to our individual decisions is also valid for defining safety requirements in technical standards. Consequently, also medical devices are not requested to be safe in the original meaning of the term and consequently to be "free of any risk". Even for medical devices (only) the ratio must be reasonable between technical-economic effort and achieved safety benefit.

2.3.1 Essential requirements

The European medical devices directive and national medical device laws define legally binding protection goals which must be met by every medical device /7/, /14/, /53/.

1. Acceptable risk/benefit ratio

Even in medical technology the required safety level is the result of a social compromise between acceptable costs and achievable increase of safety. Therefore, the obligatory essential requirement is (only) that medical devices must not have *unacceptable* risks when weighed against the benefits to the patients, when used under the conditions and for the purpose intended. Risks from human error, mistake or insufficient knowledge and experience should also be considered.

This general objective leads to important questions. How and from what can ***benefit*** be measured and made quantifiable for comparative assessment? In addition: what is benefit related to? To rapid healing? Or – if this is impossible – to improved quality of life? Probably only to alleviation of disease – or to increased duration of life, how painful or troublesome it may be? Or is benefit just seen economically, how quickly and cheaply is it possible to release the patient again to home care?

On the other hand, it needs to be clarified, how ***risk*** can be determined and quantified (see Sect. 2.2.2). Finally, the question is how small must a risk be to be "acceptable"? And above all, who decides what is acceptable and what not? Since individual and social risk perception is subjective, the same risks might be assessed differently in different countries and regions (see Sect. 2.1.1). To account for this the 3rd edition of the generic medical device standard /27/ has reduced several specified safety requirements and shifted responsibility of implementation to manufacturers. Now, they have to derive their decisions from their risk management process and implement safety precautions – if they consider it necessary.

2a. Accounting for the generally acknowledged state of the art

Design and construction of medical devices must account for the generally acknowledged state of the art of science and technology. This means that even in technology it is not required to immediately implement all new feasible solutions. The question what should be considered the generally acknowledged state of the art is crucial. In general, it comprises the status described in accepted technical standards and the safety goals defined therein. In addition, the European Commission may declare safety standards harmonized with the directive which means that meeting a harmonized standard is considered meeting the particular essential requirement.

Therefore, electromedical devices must meet the following safety goals:

- as usual also in general technology, medical devices must provide (only) ***double protection*** (and not 3-fold, 4-fold or 5-fold). This means that in case one protective means fails another redundant means of protection must be available which still provides the equivalent degree of protection. Like in other fields of technology this principle relies on the assumption that simultaneous appearance of two independent failures has such a low probability that it is not reasonable to demand for full protection also in that case.
- safety (in terms of an acceptable risk/benefit ratio) shall be provided (only) under the following conditions
  - during intended use (which is defined by the manufacturer in the instructions for use);
  - under the intended conditions for installation, supply with energy, cooling media (e.g. water), supporting media (e.g. compressed air), climate, electromagnetic environment (e.g. low background fields for biosignal recording) as defined by the manufacturer in the instructions for use;
  - with performed maintenance (inspection, service and periodic testing) as defined by the manufacturer;

- during the expectable or intended lifetime of the device (defined by the manufacturer in the instructions for use);
- at reasonably foreseeable errors;
- at reasonably foreseeable misuse;
- when used with intended knowledge (as addressed by the instructions for use).

2b. Integrated safety

Design and construction of medical devices must follow the principles of "integrated safety." This means that manufacturers are not completely free in selecting protective means. In fact, there are various options for safety precautions which can differ considerably in regard to their efficiency. The obligation to integrated safety means that in principle the most efficient measures must be preferred over less efficient ones. Depending on efficiency the following safety levels are differentiated:

- ***inherent safety*** achieved by safe design to avoid, reduce or minimize risk. This option must be chosen wherever it is possible and economically reasonable. The justified effort increases with the degree of inherent risks. Inherent safe design comprises insulating voltages, monitoring and limiting overheating, covering rotating parts, limiting output parameters to safe levels etc.
- ***indirect safety*** is achieved by auxiliary protective measures. They are acceptable if constructive means are not possible, reasonable or do not make sense. For example, while stray X-radiation can be shielded, emission of X-rays intended for diagnostic or therapeutic application is essential and cannot be prevented. Therefore, protection must be achieved indirectly such as by limiting access to X-ray rooms (assuming that persons entering are only those who have special training and are aware of the specific risks), key switches to limit use to authorized persons, personal protective means, increasing concentration levels by requested confirmation through an additional action, alarms for dangerous outputs (e. g. X-ray, laser, RF surgical currents) or when exceeding safe levels (e. g. when muscle stimulating currents exceed letgo thresholds) etc.
-  ***indicative safety*** by warnings only is least efficient. It is permitted only if more efficient approaches are not possible or reasonable (see Sect. 2.2). Examples for indicative safety are prohibiting reuse (single-use products), warning against unfavourable conditions (e. g. "Do not directly expose to sunlight!"), informing on installation requirements (e. g. "Only connect to installations for medically used rooms!"), giving instructions for use ("Not intended for explosive environment!") or transport ("Do not tilt!").

If the use of a medical device is assiociated with inherent risks, the instructions for use must contain all relevant warnings, and the device must be marked with a symbol demanding the user read the instructions for use.

3. Achieve intended performance

Medical devices must achieve the performance intended by the manufacturer and/or claimed in advertisements. This must be demonstrated by a clinical assessment file

(Chap. 3.2). This essential requirement is not trivial. It can trigger complex and expensive investigations such as clinical studies in particular in case of novel and innovative devices. This should prevent “miracle products” from being put on the market. Even if such products would not be harmful on their own they may be risky because of dangerously delaying application of efficient methods. This can hinder or even prevent healing.

4. Withstand conditions of use

Medical devices must be designed and constructed in such a way that for their whole lifetime their characteristics and performance are not affected to an unacceptable degree when subjected to stress occurring during normal conditions of intended use. To fulfil this requirement and account for ageing and abrasion, the manufacturer may shift duties to the operator (in terms of periodic maintenance and recurrent testing) and/or limit the intended service lifetime of the device.

5. Withstand storage and transport

Medial devices must perform as intended when handed over to the client and must be designed, constructed and packed so as to assure that it will not be adversely affected during transport and storage. For this reason the (transport) packaging must be properly designed. If necessary, special conditions for transport and storage must be defined. The importance of this requirement increases with length of transport and exposures of devices, for example it they are shipped and/or transported into other climate regions.

6. Unintended side-effects

Medical devices must not have unintended side-effects associated with unacceptable risks. However, unintended side-effects may not be avoidable. For instance, during muscle stimulation and RF surgery electric currents could cause burns at electrodes, infusion pumps could deliver air bubbles into the blood vessel, the cuff of external blood pressure measuring devices could impair blood perfusion of the distal extremity, an endoscope could cause bleeding and agglomeration of erythrocytes which consequently could cause thromboses. Unintended side-effects could also appear after considerable delay, such as for breast cancer caused by (former) silicon breast implants. Therefore, clinical assessment must include unintended side-effects, clinical studies might be necessary for clarification and post-market surveillance should be able to identify long-term risks (see Chap. 3.2).

7. Sufficient information

Each medical device must be accompanied by all information needed for safe use. This information may be on the product (e. g. short instructions for defibrillation, type label), on the packaging or in the instructions for use. Manufacturers have to consider the skills and knowledge of the intended users. For home-use medical devices adequate presentation and wording is a particular challenge. In case of lay application the information should be given at an intellectual level at which a 10-year-old child (with elementary school education) can understand and follow the information /32/.

Example:

... if mode wish, with conect select, look device offswitch. Attention conect plug. Side let in cause damage circuit possible. Function act in fact by jumper bring away ...

Information (and device labelling) must be written in an acceptable language (in Germany and Austria the German language is mandatory). Translations into other languages have to be checked for correctness and understandability (e. g. by involving a native speaker). Deterrent examples (such as that given in the text box) demonstrate that machine-based translations may be absolutely inadequate.

This essential requirement also obliges manufacturers to provide information on intended maintenance and recurrent testing including test intervals. Leaving this information out with the remark that this kind of service must be carried out by the manufacturer's own staff is not allowed.

8. Constructive requirements

There is a series of further essential requirements concerning design and construction of medical devices that are listed in Annex I of the medical devices directive /7/, /14/. They address general safety goals in regard to specific risks such as from

- chemical, physical and biological properties of materials with particular attention given to emission of substances that may be toxic or carcinogenic, bioactive (medicinal products), have animal origin or are derived from human blood.
- infection or microbial contamination with particular attention given to tissues originating from animals, and sterilization.
- construction and environmental properties.
- measuring functions.
- unintended or intended radiation with particular attention to ionizing radiation.
- internal and/or external energy sources.
- mechanical and thermal stress.
- delivery of energy and/or substances.

Detailed requirements are contained in the generic standard EN IEC 60601-1 and, if applicable in special parts 2 devoted to specific types of devices such as RF surgery (EN IEC 60601-1-2), nerve and muscle stimulators (EN IEC 60601-2-10), infusion pumps (EN IEC 60601-2-24) or magnetic resonance imaging devices (EN IEC 60601-2-33).

2.3.2 Fault conditions

Medical devices must provide sufficient protection under normal and single fault condition. This leads to the question what is understood by a single fault condition?

Single fault

A single fault is any hazardous situation that needs to be taken into account, but whose probability of occurrence is low enough to allow independent consideration. However, if a single fault causes a consecutive "single fault," both failures are considered as one single fault (e. g. breaking of a safety chain in case of a fault of the mechanical fixation).

Examples of single faults are

- the ***failure of a protective measure*** (e. g. damage to electric insulation, interruption of the protective earth conductor, failure of a temperature limiter, a movement limit switch or a gasket);
- occurrence of one ***abnormal condition*** (e. g. defect of an electronic or mechanical component, overload of an electric circuit, leakage at liquid or gas connections, impairment of cooling, blocking of motors and ventilators);
- ***human errors*** (e. g. *mishaps* such as spilling of liquids, dropping of hand-held applied parts, exceeding specified duration of use; *unintended actions* such as activating an actuator, disconnecting interconnections; *mistakes* such as confusion of connectors or control elements; *misuse* such as ignorance of extensive checklists, insufficient disinfection because of difficulty of demounting parts);

No "single faults"

If a hazardous condition occurs too frequently, it is considered to be "normal" and thus needs protection by two independent means. No single faults are

- ***frequent fault conditions,*** e. g. exhausted batteries, short-circuiting or free running of electric stimulating electrodes, pulling on patient connections (electric conductors or valves);
- ***insufficiently designed protective measures,*** e. g. insulation thickness, air distances, creepage distances.

Remark: Insufficiently designed protective means are assessed as being not existent at all.

Single-fault safe

Medical devices have to be free of unacceptable risks during their whole expected service life under intended conditions and single fault conditions (single-fault safety). This is fulfilled

- if in case of a single fault an equivalent second protective measure is available, and if the single fault can be detected right before a second protective means fails or if another single fault occurs. This can be achieved for example by protective earthing causing short-circuiting and switching off of the electric circuit; by providing second insulation together with periodic testing; by main mechanical fixation and an additional safety chain;

- if in case of a single fault the probability of failure of the second equivalent protective means is negligible during the whole expected service life;
- if a single protective means is provided that has a negligible probability to fail during the expected service life (e. g. reinforced insulation, components with high-integrity characteristics, suspended mass with overdesigned (e. g. 8-fold) safety factor.

2.3.3 Safety concept

To achieve the safety goal during the whole service life of a medical device the safety concept relies on three parties: the manufacturer, the operator and the user.

1. the ***manufacturer*** is responsible for device safety. Therefore, he has to apply safe design, construction and manufacturing. However, to be able to cope with this responsibility for the whole expected service life he has to delegate tasks to users and operators by including duties into the instructions for use both in regard to application as well as to maintenance.
2. the ***operator*** is obliged to maintain medical devices according to the specifications of manufacturers both in regard to maintenance and to recurrent testing in intervals and to an extent as specified by the manufacturer.
3. the ***user*** is obliged to apply the device according to the instructions given by the manufacturer and with the accessories specified. This requires user's knowledge of the instructions for use and training. In addition to that, prior to each new application users are obliged to visually check that the device is still in order.

Remark: In Austria users have to be verifiably trained in the use of the various types of medical devices (not just kinds of devices) and be informed on their specific risks. Only after this are they entitled to use these devices. Hospitals have to keep individual records of this training (personal "device driving licences").

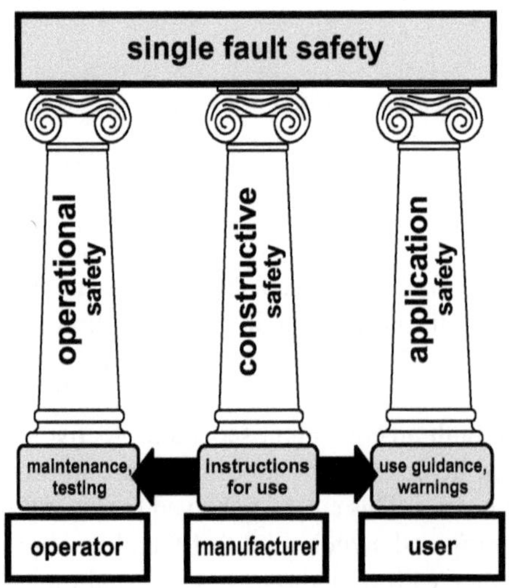

Figure 2-16: The three-column safety concept in medical technology

3 Application safety

3.1 Usability

During application of medical devices risks due to human error cannot be excluded. These may be due to the special situation (e. g. due to emergency situations in operating theatres, intensive care units or ambulances), insufficient information (e. g. due to shift changeover), overload (e. g. because of stress, emergency, unexpected events), inattentiveness (e. g. because of distraction or fatigue at the end of a shift).

To minimize risks from false reactions, misunderstanding or mistakes medical devices must meet usability requirements /7/, /29/. Manufacturers must implement a process allowing detection, analysis, control, avoidance and/or minimization of such risks by adequate design preventing user errors provoked by foreseeable misuse or mistakes even under foreseeable critical conditions (EN 60601-1-6 /29/). As an example, it can be foreseen that lengthy instructions for defibrillation might not be read in an acute emergency, and for this reason, advice must be given by other means such as by a sequence of images on the device, or by semiautomatic solutions, with step-by step acoustic advice generated by the defibrillator.

Another foreseeable situation would be the reuse of (expensive) single use devices. Even if this is in contradiction with the intended use, now a manufacturer must explicitly warn of associated hazards and has to list reasons in the instructions for use. An example of such a warning is given below:

> ***Warning!*** *The device is for single use only! Do not reuse, reprocess or resterilize. Refurbishment may compromise the structural integrity of the device and/or lead to device failure which, in turn, may result in severe patient injury, illness or death. Mechanical degradation of the device's surface may lead to inefficient disinfection/resterilization and consequently to microbial contamination. This may cause patient infection or cross-infection and consequently illness or death of patients!*

> *Remark: In contrast to repair refurbishing single-use devices is considered a (new) manufacturing process. Refurbishers are considered to be the (new) manufacturer with all obligations and have to assure that refurbished devices meet the essential requirements. Consequently, they have to again CE-mark refurbished devices.*

False actions can be either performing or omitting required operations. If such risks cannot be controlled by design, depending on risk analysis alarms and interruption of operation, even breakdown (*fail-safe*) could be accepted, if it does not itself lead to intolerable risk (Figure 3-1).

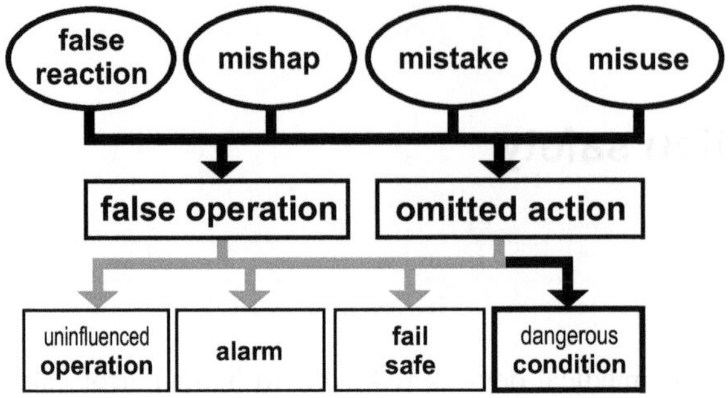

Figure 3-1: Foreseeable human errors and potential consequences of false operation and/or omitted required actions

Examples of foreseeable critical situations are:

- spatial conditions (e.g. hospital, ambulance, stretcher, home);
- social aspects (e.g. teamwork, shift changeover, split responsibility, present family members and children);
- technical realization (e.g. too close spacing of push-buttons, similar-looking icons, interchangeable connectors, complex operation, processes that are difficult to understand; interaction with other devices, demanding preparation, expensive maintenance, complex installation);
- hygienic aspects (e.g. demanding disassembly, complex preparation for disinfection or sterilization);
- physical condition (e.g. illumination, air pressure, temperature, humidity, weather, altitude);
- mental condition (e.g. stress, overworked, too demanding, tiredness, surprise, startled);
- human shortcoming (e.g. distraction, absent-mindedness, laxity, ignorance, flippancy).

Therefore, manufacturers have to pay attention to

- placement of operation elements (e.g. clear-cut and logical operation sequence such as selecting output and delivered energy);
- sufficient spacing (e.g. separation of contrary operations such as activation and deactivation);
- configuration of operation elements (e.g. size of buttons and display);
- work flow (e.g. logical operation such as increasing output by turning clockwise, no unusual sequences);
- failure tolerance (e.g. no excessive consequences of failures such as explosions from (frequently chosen but actually forbidden) use of alcohol, no severe tissue burns following careless application of electrodes);
- quick operational readiness (e. g. short checklists);
- adequate display (e.g. sufficiently long display of critical situations or error messages);

- clear, unambiguous and targeted information (instructions, symbols, displays, wording and content of instructions for use);
- sufficient tolerances (e. g. mechanical tolerances of connecting elements).

Usability is especially important for home-use devices. Plain wording and explanations without precondition of special knowledge and training, avoiding lingo and special symbols are imperative. Easy handling and avoidance of dangerous output as well as an adequate strategy for maintenance and recurrent testing are essential /32/.

3.2 Clinical assessment

It is one of the essential requirements that all medical devices irrespective of their conformity class must in fact have the intended and/or claimed medical performance. However, they must not be associated with unacceptable unintended side-effects either. Manufacturers must demonstrate this by clinical assessment (Figure 3-2), but they may do this by referring to the existing state of knowledge such as by

- market experience with other comparable products;

  Remark: Reference to experience with other comparable products is accepted as an indicator for proved clinical efficiency. However, in view to the fact that the belief in the efficiency of an objectively inefficient method could lead also to beneficial effects (placebo-effect) makes this kind of evidence weak, if it is not supported by other indicators such as plausible interaction mechanisms.

- acknowledged scientific literature (rather than grey literature of questionable seriousness);
- published and/or unpublished reports of sufficient depth and quality;
- other documented clinical experience;
- results of clinical studies with other comparable devices.

Reliability and validity of such information must increase with the suspected inherent risk of the assessed device. If evidence from these sources is not sufficient, not comparable or not applicable because of different methodology, characteristics, performance, site of application, medical indication and/or used material it might be necessary to clarify open issues by performing a clinical study.

Remark: Clinical assessment must be performed and documented for all medical devices irrespective of their conformity class. However, third party approval by a notified body is only required for class IIb and class III devices. In fact, there are already a series of medical products on the market (with low inherent risk) whose performance is in doubt. They are tolerated as long as their use is not associated with elevated risk. An example of such devices are bioresonance devices which register biosignals, partly invert them by a top-secret company-specific method and feed them back into the body via the same electrodes. It is claimed that by this procedure "bad oscillations" are converted into "good" ones and patients will get rid of several illnesses.

Clinical study

Clinical studies are not only costly, laborious and time-consuming (Figure 3-2). In addition, they are regulated by strict requirements (Annex X MDD /7/, /14/, /53/). Prior starting manufacturers must have already assessed the conformity of their device with all essential requirements – except those that are to be checked by the study, and write

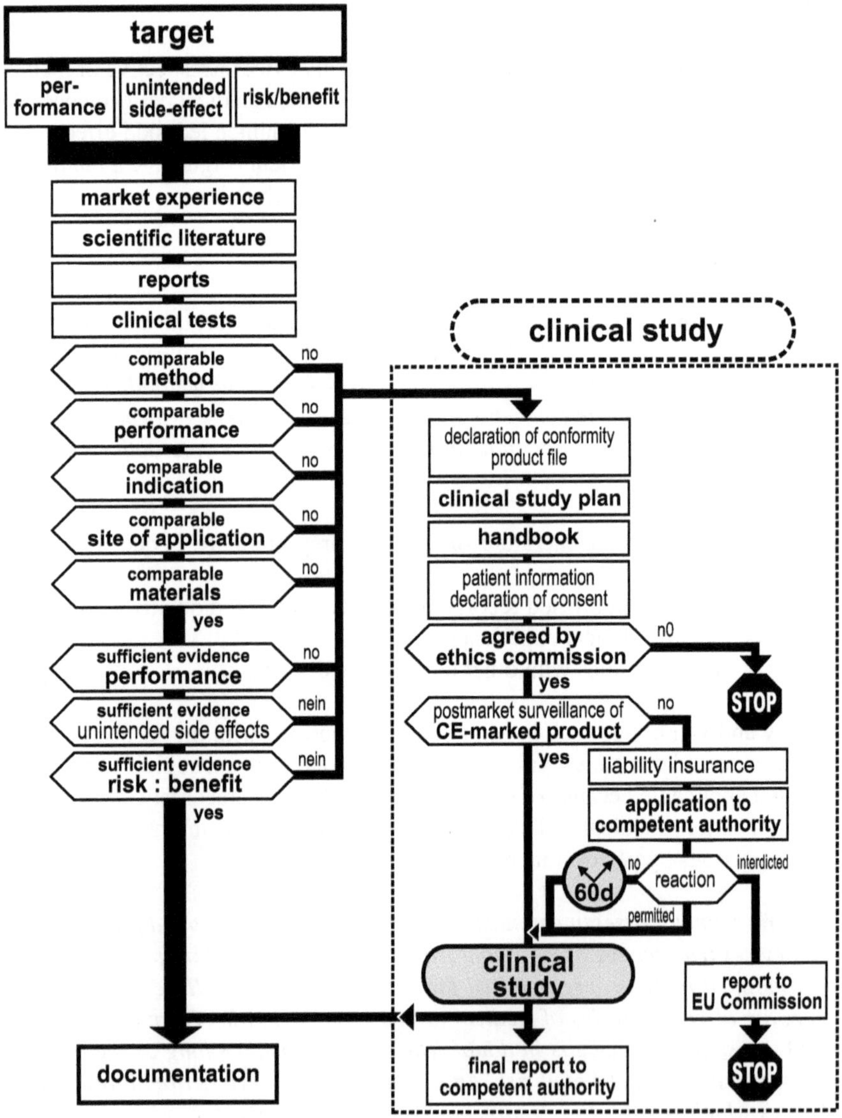

Figure 3-2: Flow chart of clinical assessment of a medical device

a related declaration according to Annex VIII MDD. They have to elaborate a clinical study plan and a handbook for clinical testers, information for participants, and have to base their study on informed consent of participating patients. These documents must be presented to and agreed upon by an ethics commission and then be forwarded to the competent authority for permission. If it is not interdicted within 60 days (or permitted earlier) the study might be started. However, further detailed requirements as contained in the medical devices directive must be met.

Remark: A systematic post-market clinical surveillance study of already CE-marked products is also considered a clinical study and must be approved by an ethics committee. However, it is not required to be reported to the competent authority and does not need approval by it.

4 Biocompatibility

Basically, any physical contact of a body with material is associated with diffusion and more or less pronounced exchange of molecules across the contact area. Health risks may occur if health-relevant bioactive substances are delivered to the body. This could have different adverse consequences (Figure 4-1), such as

- eliciting allergies in terms of overreactions of the body's own immune system by producing antibodies against normally non-critical substances such as nickel (e.g. spectacle frames) or latex (e.g. surgical gloves);
- tissue inflammation;
- poisoning (toxicity);
- initiating cancer (carcinogenicity);
- enhancing malignancy of existing tumours (tumour promotion);
- causing malformation of foetuses (teratogenicity);
- causing abortion.

Manufacturers must assess and assure biocompatibility of contact materials in particular – but not restricted to – of applied parts. For this purpose, it would be necessary to know the composition of materials and to assess substances in regard to the intended use of the device. For instance, plastic such as PVC (polyvinylchloride) usually contains additional substances to achieve the properties required in medical device technology such as elasticity, stiffness and fire resistance. Many of these additives are adverse to health and are toxic, abortive, teratogenic and/or carcinogenic. As an example, an ophthalmic surgical device had to be recalled from market because the distance holder delivered endotoxins into the cornea and led to inflammation (see text box).

Ophalmic devices called back

California: FDA requested a callback of 4,339 ophthalmic surgery devices because of a bio-incompatible applied part. During normal use enhanced amounts of endotoxins had diffused into corneas and caused post-surgical eye inflammations.

To assess biocompatibility the following parameters have to be considered:

- the accumulated (!) ***exposure time***. In contrast to the uninterrupted exposure time as used to decide upon the conformity class of a medical device, for assessing biocompatibility the exposure time is summed over the entire contact duration, for example for the whole intended treatment procedure and not just for one single treatment. Exposure time is classified as short-term (<24 hours), long-term (1–30 days) and continuous (more than 30 days). As an example, the accumulated contact time of an individual surgical glove can be classified as short-term. However, since a surgeon would have to wear such gloves several hours almost every working day during his whole working life his accumulated exposure time is considerable longer. Therefore, for assessing biocompatibility surgical gloves are classified as "continuously" contacting the body. In fact, allergy to latex is encountered over-proportionally among surgeons.

- ***kind of contact*** for example with intact skin, with wounds or direct contact to flowing blood.
- potential ***health consequences*** of the material, (Figure 4-1) for example causing skin irritation or inflammation (e.g. endotoxins), poisoning (e.g. halogenic compounds, heavy metals), allergies (e.g. Nickel, latex), cancer (e.g. plastic softeners), foetal malformation (e.g. lead).

An important precondition for assessing biocompatibility is to know the detailed composition of materials coming into contact with the body. This refers not only to the basic material but also to its additives. This requirement is not trivial. Since such detailed information is requested mainly from medical device manufacturers only which constitute only a small market segment, material manufacturers are reluctant to analyze their material and provide such data. However, in case of insufficient information medical device manufacturers do have alternatives (Figure 4-2). These are taking advantage of experience with similar use of the chosen material, performing own (expensive) biocompatibility tests, or changing to approved alternative materials.

The European Commission's risk assessment of additives to plastics has resulted in the banning of phthalates. Such additives are used to soften PVC. They are considered carcinogenic, genotoxic and teratogenic and should be avoided where possible. Therefore, manufacturers are requested to mark their products and draw the attention of users

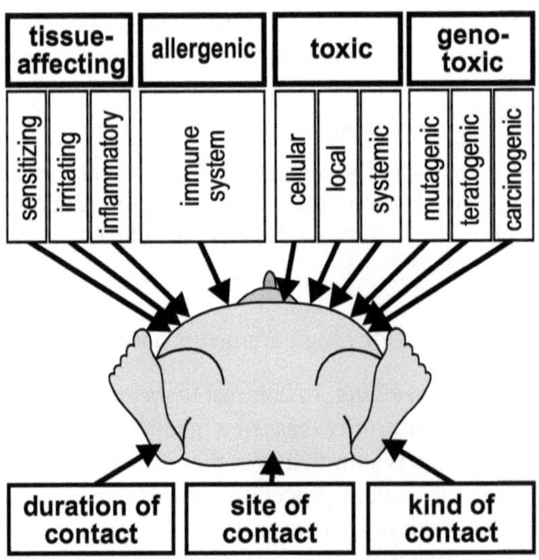

Figure 4-1: Potential biological reactions to bioactive substances

to the content of phthalates. Particular attention must be given to products of PVC or (PET) (polyethylenterephthalate) contacting fluids that are intended to be re-infused into the body (e. g. infusion bags, infusion sets, blood bags, catheters). If pregnant women or children are treated with phthalate-containing products this is acceptable only in justified exceptions /7/.

! Phthalate-containing products must be marked

Remark: Alternatives to soft-PVC dependent on the required properties are softener-free polyethylene (PE), polypropylene (PP) or polystyrole.

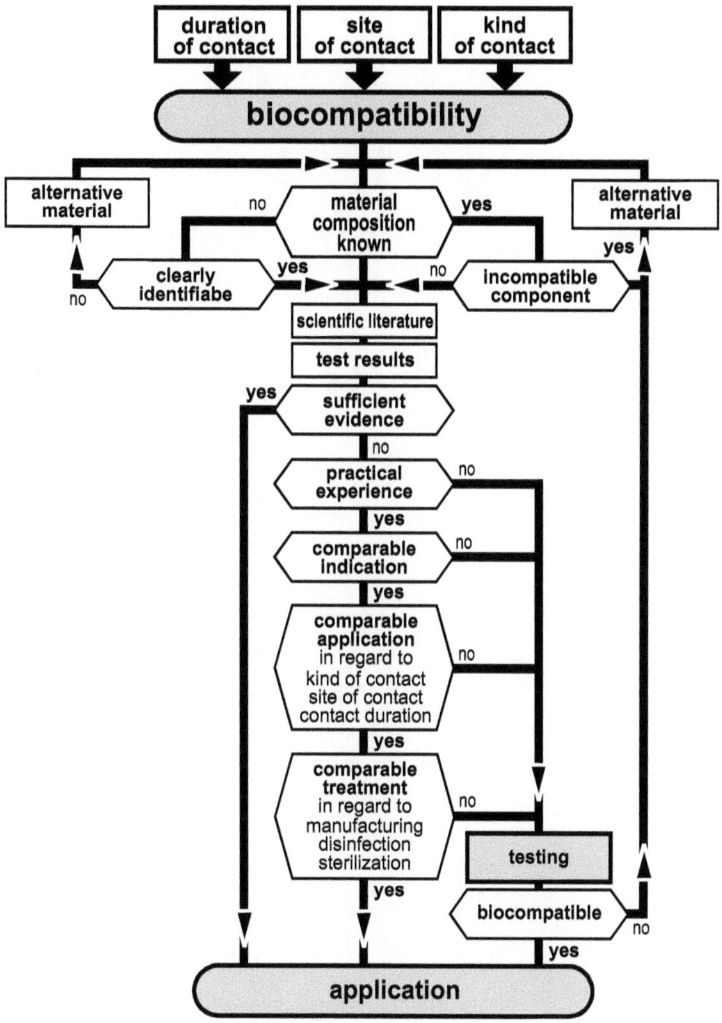

Figure 4-2: Flow chart of biocompatibility assessment

5 Hygiene

Medical devices have to be designed so as to minimize as much as possible infection risks to patients, users and others including clinical engineers. Medical devices with high infection risks are those that are intended to come into contact with body fluids (e.g. endoscopes, suction devices, dialysis devices). A particular hygienic problem is germ proliferation. In devices containing residual liquids germs might proliferate very quickly. Because of their exponential increase germs quickly reach enormous numbers. As an example, in the case of a 20-min duplication time, within 8 hours one germ proliferates into already 2.10^7 and after one day even 5.10^{21} germs.

Experience shows that also medical technicians might be exposed to infection risks, for example during periodic testing or if they were given a device for maintenance or repair without preceding disinfection. Therefore, it is important to be aware of this risk and to clarify the origin and hygienic pre-treatment of devices before starting work.

! Device maintenance and repair only after clarifying the hygienic status

Therefore, devices or applied parts at risk of becoming contaminated must be designed so as to be able to be cleaned, disinfected and/or sterilized. For this purpose they have to withstand mechanical forces as well as chemical and/or thermal stress.

Chemicals are required for reliable *cleaning* (in addition to mechanical treatment to dissolve fat and proteins) as well as for *disinfection* or chemical *sterilization* to inactivate and kill germs. However, apart from their concentration their efficiency depends both on temperature as well as on exposure time (Figure 5-1).

Mechanical cleaning, for example by brushes, might considerably affect applied parts and roughen their surface. Chemicals could accelerate a material's ageing and corrosion and change its properties such as critically reducing its elasticity. Thermal (over)stress could also accelerate material ageing, lead to deformation and embrittlement, or even damage such as impairing or even erasing the piezoelectric effect of ultrasound transducers.

Therefore, in their risk analysis manufacturers must assess the design and construction of a medical device and its material in regard to potential impact of and/or suitability for cleaning, disinfection and/or sterilization and must define appropriate methods and substances to minimize hygienic hazards and impairment of the device.

The choice of materials and specification methods intended for cleaning, disinfection and/or sterilization influence device ageing and hygienic risks. Therefore, adequate design is important. It involves surface structure and shaping as well as construction details such as accessibility and disassembling of parts.

Infection risks increase with rough and structured surfaces, grids, openings and grooves, and with the required effort to get access to parts and/or disassemble devices for disinfection or sterilization. Therefore, for instance, for hygienic reasons membrane keyboards are preferred over conventional keyboards.

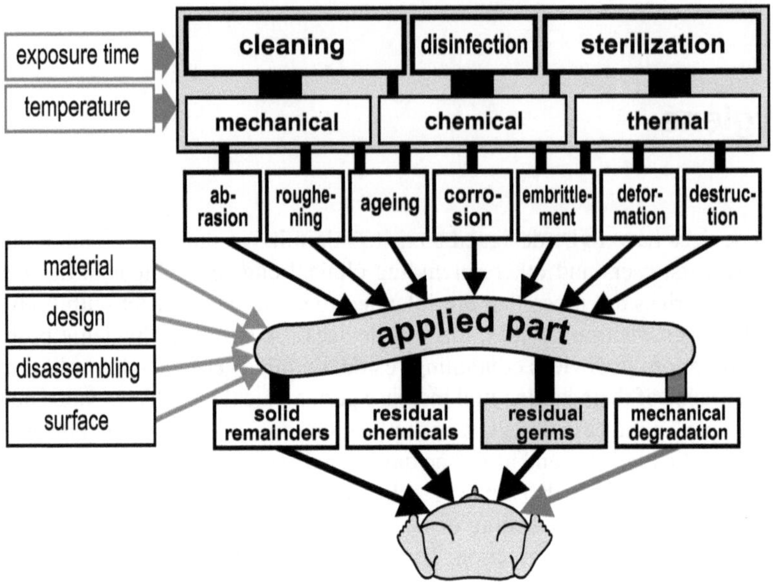

Figure 5-1: Hygienic aspects of medical devices

6 Environmental safety

Medical devices may interact with the environment in two different ways: Depending on their kind, construction and function they may be influenced ***by*** the environment in many ways as well as having an impact ***on*** the environment by adversely influencing or even endangering it.

6.1 Interference with the environment

6.1.1 Environmental conditions

In general, electromedical devices are designed so as to operate as intended under the following conditions /27/:

environmental temperature	+10 to +40°C
relative air humidity	30 to 75% rH

Depending on the intended use it might be necessary to enlarge or restrict these limits. As an example, defibrillators which may be used in an emergency also in open air or bad weather conditions need temperature limits extended down to 0°C and increased air humidity to 95% rH. In addition, they have to have an increased protection against ingress of liquids (IPX1, see Chap. 9.2.1). Home use devices have the temperature range extended to 5°C and an extended humidity range (15–95%). In addition, they require an increase protection against ingress of liquids (in general IPX1, for movable devices IPX2) /32/.

6.1.2 Electric installation

Electric installations for medical locations differ from general electric installations in regard to several aspects. The safety measure "protective earthing" (SC 1) critically depends on additional requirements of the electric installation (Chap. 8.5.1). In medically used rooms it is assumed that the skin no longer protects the patient. Therefore, electric installation in medical locations must provide an increased degree of electric shock protection. In addition, the patient's dependency on life-supporting devices may require supplying power with increased reliability compared to general rooms. In addition, it might not be acceptable to provide protection during single fault conditions by switching-off electric circuits and, hence, deactivating life-supporting devices. In addition, the kind and characteristics of electromedical devices may require specific properties of the installations such as low internal resistance of the power source to minimize voltage drops during high-power short-time activation of X-ray generators, or uninterrupted power supply (e. g. for gamma cameras to avoid damage of the large scintillation crystal by thermal stress, or for cooling systems of superconducting magnetic field coils of magnetic resonance imagers to avoid heating above transition temperature with the risk of explosive expansion of liquid cooling media Helium and Nitrogen).

For these reasons compared to households the requirements for electric installations in medical locations differ in regard to increased electric shock protection, availability and reliability.

1. increased ***electric shock protection*** must account for a patient's increased vulnerability due to lacking protection from the skin. This is achieved by adequately designing protective means so as to limit touch voltages during single fault condition to 25 V_{AC} or 60 V_{DC}, respectively (see Chap. 8.2). In addition to installed overload circuit breakers, earthed power supply systems require residual current circuit breakers. These devices monitor electric currents flowing to and from electric loads and interrupt the circuit sufficiently quickly if due to an insulation failure currents choose another route back such as across the patient. In medical locations residual current circuit breakers should be installed which interrupt circuits within 0.25 s at a nominal residual current of 30 mA.
2. we know that electric blackouts can occur even for longer time periods for different reasons such as thunderstorms, hurricanes, earthquakes or internal breakdown (see text boxes). In general, in households apart from inconvenience this may not have dangerous consequences. On the contrary, after some delay this might even lead to increased birth rates! However, in vulnerable medical locations blackouts might be life-threatening to patients and, hence, cannot be tolerated. Therefore, hospitals must be able to sustain at least safety-relevant operation through its own means. Therefore, increased reliability of power supply is provided by two different autonomous electric safety sources:
   A slower ***safety emergency power supply*** system is based on diesel generators providing power within less than 15 s. However, even 15 s changeover time might be too long such as in critical phases of surgery where immediate action might be required.
   Therefore, in addition, an ***auxiliary emergency power supply*** system is foreseen based on batteries providing power with a changeover time of less than 0.5 s. It is intended to supply the most important appliances such the operation lamps or RF surgery devices.
   However, voltage drops during changeover can have unintended side-effects. They could reset software-controlled devices to default settings such as infusion pumps with consequential stress for the staff, for example of intensive care units, as they scramble to reset devices to work as initially intended. As an alternative, uninterruptible power supply systems provide electric power without any changeover time thus avoiding critical voltage drops and software problems.

New York, August 2003: System overload of the Niagara Mohawk utility caused a wide-ranging blackout: 50 Billion inhabitants of New York were without electricity for 18 hours, surrounding areas had to wait for electricity for days.

Bonn, November 2006: In Northern Germany a planned power line disconnection by E.ON triggered a chain reaction. Ten Billion people in Europe had to cope for several days without electricity.

Chemnitz, January 2007: Windstorm Kyrill mowed down power line masts in central Europe. Hundreds of Thousands of people were without electricity for several days. Reconstruction of power lines took months.

Remark: To avoid breakdown while restoring power supply an excessive initial load must be avoided. This is done by sequentially reconnecting appliances to emergency power supply systems according to their importance for the patient and hospital operation.

In those locations where not all socket outlets are connected to an emergency power supply (e.g. surgical ambulances) it is important to clearly mark differently supplied socket outlets. This can be done by labelling or by colour-coding. Frequently, green is used to signal connection to an emergency power supply and orange to indicate connection to an auxiliary emergency power supply. Medical staff should be aware of the different meanings and connect devices accordingly. Already legendary is the answer of experienced medical staff to the question why some socket outlets had different colours. They just shrug their shoulders replying "because of the wish of the architect." Obviously, this dangerous wrong answer showed that they were a considerable risk factor themselves. The reason is because it is not irrelevant to know which socket outlets would continue providing power in any situation. Only this allows continuing operation of life-sustaining devices or quickly adequately changing connections in case of a blackout.

! Emergency power-supplied socket outlets must be marked – and known

In addition, it is also important to know which socket outlets belong to which electric circuits. Only this enables staff to react properly for example in the case when the installation circuit breaker switches off a circuit to protect from overload. Since it is required to provide socket outlets connected to another redundant electric circuit within the patient environment /55/, it would be important to be aware that plugs of life-sustaining devices could be easily changed to nearby socket outlets of the redundant circuit which are still supplied.

! Socket outlets must be marked with the assigned electric circuit

3. it has already been mentioned that double protection can be achieved by different approaches. One of them, protective earthing, relies on short-circuiting in case of insulation failure with consequential circuit interruption by the installation circuit

breaker and termination of the failure situation. However, in an operating theatre this protection principle would be hazardous. The reason is that in single fault condition all other medical devices connected to the same electric circuit would also be switched off, including life-supporting devices. This could indirectly be hazardous to the patient and would not be acceptable. The way out of this dilemma of meeting both requirements, namely double protection and reliable power supply, is offered by an (expensive) principle, namely, insulating ***all*** live parts of an electric circuit from earth. This is done by generating an earth-free insulated power supply system (IT-power supply). If now under a single fault condition one pole of the current source contacts the earthed device enclosure … nothing happens, no short-circuiting, no interruption of the circuit, nothing! Simply because the IT current source is now just connected to earth but the electric failure circuit remains open with no failure current flowing (Figure 6-1). The only thing that occurs is that the insulated power supply (IT-) system changes to a grounded (TT-) system – which is already common for general power supply – but operation can be continued without any disturbance.

Remark: Only, if a second independent failure occurred and, as an additional requirement (!), if this would affect the counterpole of the circuit, something would happen which would be common also in other locations: Now the second failure would cause a short-circuit, the installation circuit breaker would be activated and the circuit interrupted.

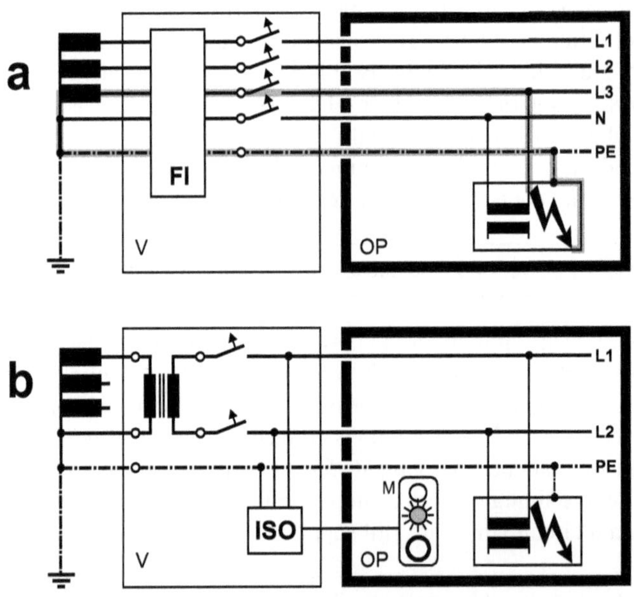

Figure 6-1: Consequences of an insulation failure (a) in an earthed (TN-) power supply system with a short-circuit current (grey) flowing and (b) in an isolated (IT-) power supply system with activated alarm of the insulation monitoring device (grey). V … distribution box, OP … operating theatre, F … insulation failure, ISO … insulation monitor, M … display panel, L1, L2, L3 … live conductors, N … neutral conductor, PE … protective earth conductor

Allowing there to be no effect from the single fault condition has one downside; the single fault condition is no longer detectable, and would persist until the following safety inspection (if there is one). Therefore, another means is required to detect the problem and alarm users to this situation. This is the application of an insulation monitoring device. It continuously monitors the insulating resistance between all live parts and earth potential. It is situated in the distribution box and connected to a display panel in the room. If an insulation failure occurs, an acoustic and optical alarm is issued (the acoustic alarm can be prompted only), but there is still no reason to panic. The surgical team can carry on working without any safety problem. At the end the failure can be located and eliminated without any stress.

Of course, power system reliability must be paid for. In this case, additional expenses result from an extra safety transformer. It is usually situated in the distribution box and must be designed to be large enough to supply the required electric power. To check for the correct operation of the insulation monitoring device, the display panel contains a test button. If activated, a single fault is simulated by connecting one live conductor to earth by a low resistance. Such a periodic test should be performed by the users. It is not only intended to check the insulation monitoring device but also functioning of (red and green) indicator lamps on the display panel.

! The insulation monitoring device should be checked periodically by the user

Installation concept

Additional measures for electric installation in medical locations cost additional money. Economically, it is not feasible to provide such expensive power supply systems throughout a hospital. Therefore, rooms in medical locations are classified according to their installation demands into four groups /55/, /60/:

- ***not medically used rooms*** within medical locations (e.g. tea kitchen, lavatories, lounges). In this group conventional electric installation is sufficient.
- ***group 0 rooms*** comprise medically used rooms without any special safety requirements to the electric installation (e.g. ward rooms, doctor's practice). In these rooms patient safety is not dependent on the electric installation because there is no intended use of electromedical devices or only inherently safe devices are used such as safety class II devices, or devices with internal electric current source. However, use of (protectively earthed) safety class I electromedical devices is not allowed.
- ***group 1 rooms*** comprise medical locations where also safety class I devices may be used: Therefore, safety of the patient is dependent on the increased electric shock protection of the electric installation (reduced touch voltage). In addition, power supply reliability and, consequently, emergency power supply is relevant as a back-up in case of blackouts, but protection by switching off is acceptable. Therefore, electric circuits may be interrupted in single fault condition without harm to the patient, (e.g. surgical ambulance, intensive diagnostic room, biosignal recording, endoscopy, X-ray imaging, hemodialysis rooms).
- ***group 2 rooms*** comprise the most critical locations where life-sustaining and/or life-supporting devices are used and critical surgical operations may be performed.

Patient's safety is dependent on increased electric shock protection by the electric installation and reliability and availability of power supply. For blackouts emergency power supply and auxiliary emergency power supply must be provided. Protection from electric shock by switching off electric circuits is not acceptable. Therefore, power supply must be performed by insulated (IT-) systems. Only high-power devices such as X-ray generators might be supplied by an earthed (TN-) system.

6.1.3 Electrostatic discharges

In daily life electrostatic discharges are omnipresent. They appear already if we comb our hair, put on our pullover or just stand up from a car seat. Discharges can be strong enough to be perceived or even cause an electric shock. The physical reason is that different materials, although electrically neutral, in general differ in regard to the density of their contained electric charges. If two electrically neutral materials touch each other, at the contact area charges are exchanged by diffusion. However, each migrated charge disturbs electric neutrality. This leads to an increasing potential difference and, hence, to an electric counterforce. Diffusion continues until equality of the driving diffusion force and the electric counterforce (Figure 6-2).

At the end, the electric conditions at the contact area are similar to those of an electric capacitor with two differently charged layers facing each other but separated from each other. Exchanging charges at the contact area is no problem as long as the separation process allows them to flow back into their initial materials thus restoring their neutrality. However, if the separation speed is higher than charge backflow velocity, migrated charges must stay in their host material which then loses its electric neutrality and remains electrically charged.

Therefore, electrostatic charging depends on the separation speed and the mobility of electric charges within materials (material's electric conductivity).

Materials are electrostatically chargeable if they exhibit poor electric conductivity (if they are insulators). This can be assumed for surfaces with a resistance above $10^9\,\Omega$,

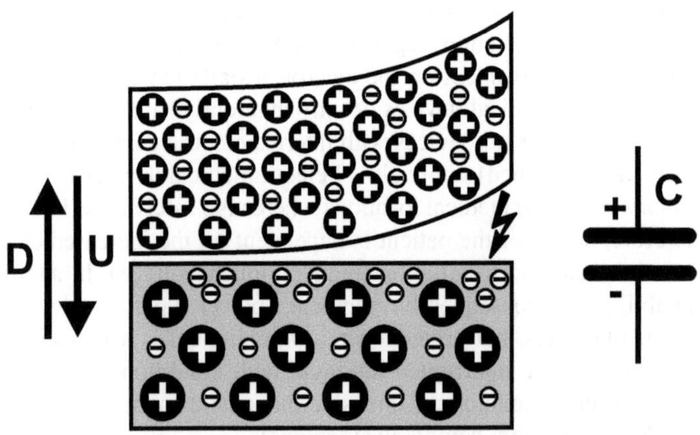

Figure 6-2: Charge exchange at the contact area of two materials until equality of diffusion force D and electrostatic counterforce U (left). Also shown is the electric equivalent diagram (right)

liquids with a specific resistance higher than 10^8 Ωm, nebulae (independent of specific resistance), and earth-free electric conducting objects (with earthing resistances above 10^4 Ω). Small objects with discharging time constants below 10 ms are considered not electrostatically chargeable /1/. Electrostatic charging should be expected on any occasion where separation involves at least one poor conducting material, for example walking over a poorly conducting floor (e. g. with PVC floor-covering). The stored charge Q=C.U and the potential discharging energy $W_E = ½C.U^2$ are determined by the charging voltage U and the electric capacity of the charged object.

The electric capacity of persons depends on their posture and their distance to the floor (Figure 6-3). Frequently it is 100–200 pF. The charging voltage can amount to 10–15 kV, but under unfavourable conditions may increase to even much higher values. Consequently, the discharging energy can become 5–23 mJ. Such energies are high enough to destroy electronic components. They are also high above the ignition energy of explosive gas mixtures. This makes persons potential critical ignition sources. Therefore, in critical medical locations precautions are required to avoid electrostatic charging (e. g. conducting floor-covers and conducting shoes).

Remark: Health risks to the patient from direct electrostatic discharging of persons to the patient's body surface can be excluded. However, health risks such as heart fibrillation from discharging of the surgeon or from medical devices (e. g. probes, sensors, tubes) to the open heart cannot be excluded.

Electromedical devices must be designed so as to exhibit minimum immunity against electrostatic discharges to the enclosure as well as to connecting conductors. Test voltages of 8 kV (air discharge) or 6 kV (contact discharge) must not cause unacceptable risks /28/. However, these requirements are not strict enough to generally exclude interference of damage from a person's electrostatic discharges. Therefore, caution is necessary in particular if device enclosures have to be removed for repair or inspection and discharging to the interior is made possible. This is because the failure level of elec-

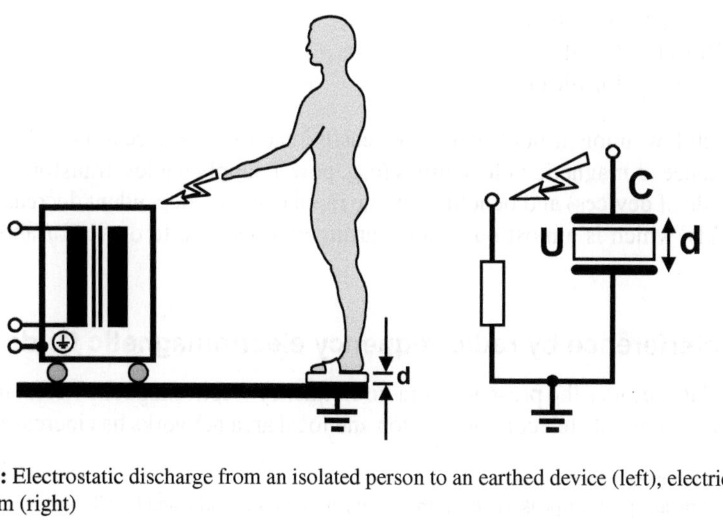

Figure 6-3: Electrostatic discharge from an isolated person to an earthed device (left), electric equivalent diagram (right)

tronic components is usually several orders of magnitude below typical discharging energies from a person.

Be careful when opening devices Electrostatic discharges may destroy electronic components

In justified cases medical devices are allowed to have reduced immunity against electrostatic discharging. In such cases devices must be marked with the standard symbol and operated only at electrostatically protected sites /28/.

6.1.4 Interference by magnetic fields

Magnetic fields are unavoidable companions of electric current flow. They may interfere with medical devices by inducing voltages into conducting loops, impair their function or even cause damage. In its close surroundings the mains transformer of a device may cause magnetic inductions of several µT (if measured with a probe of 100-cm² sensing area), and devices with electronic power control due to their frequency harmonics may cause even higher magnetic inductions.

Remark: A market survey identified electric appliances with magnetic fields up to over 8,000 μT_{rms} *in their close surroundings /47/.*

Magnetic interference fields may merit attention in particular in case of electric systems with densely packed devices. The required immunity level is defined in EN 60601-1-2 /28/ which in turn refers to EN 61000-4-8 /33/. It amounts to (only) 3.8 µT (3 A/m) /28/. This is far too low to exclude relevant interference in daily life inside and outside hospitals.

At the place of the patient much lower environmental magnetic fields may be requested such as for biosignal recording or monitoring /55/, /60/, in particular

141 nT (400 nT_{pp}[4]) for ECG,
71 nT (200 nT_{pp}) for EEG,
35 nT (100 nT_{pp}) for EMG.

Since such low ambient fields cannot be easily reached it is necessary to keep a sufficient distance to magnetic field sources (e. g. power supply cables, transformers in side and outside of devices) and benefit from the rapid distance-dependent decrease of magnetic fields, which is almost quadratic starting at a distance to device of about 10 cm /49/.

6.1.5 Interference by radiofrequency electromagnetic fields

Over the last decades the presence of radio frequency electromagnetic fields, in particular fields from mobile telecommunication and local area networks has increased consid-

[4] nTpp means peak-to-peak vaues (rather than rms values as indicated by nT).

erably. Most people already continuously carry with them small RF transmitters in terms of mobile phones emitting microwaves up to 2 W_{peak} (250 mW_{rms}) transmitting power. Even more powerful mobile sources are used by emergency medical teams.

OP-tables called back

Bavaria: Operating tables of the model 2002 had to be called back because of deficient immunity to EMI. RF surgery had activated internal motors and caused hazardous displacements of the table and the patient during surgery.

Investigations showed that electromagnetic interference (EMI) from handsets with medical devices (e.g. infusion pumps, ECG monitors) cannot be ruled out. Since thin partition walls only weakly shield microwaves, interference can even extend to neighbouring rooms. Therefore, manufacturers must analyse EMI risks and possibly request keeping a safety distance. Recommended safety distances are 1 m to handsets and several metres to emergency radio units. However, immunity against electromagnetic interference can be increased by adequate design and/or grounded electromagnetic shields inside devices. This is of particular importance for medical devices intended to be used in uncontrolled or uncontrollable environments such as homecare devices or electric wheelchairs.

Remark: To ban handsets already at the entrance to a medical location is an overreaction and counterproductive. Such large-scale bans inherently stimulate indifference even in critical places (e.g. close to or inside intensive care units or operating theatres).

6.2 Impact on the environment

6.2.1 Electromagnetic Emissions

Electromedical devices might not only be affected by electromagnetic interference but also be a source of unfavourable influences to the environment themselves. Disturbances may be due to conducted interference via power cables and/or signal input/output connections (e.g. fast transients, bursts, surges or voltage dips, variations, and interruptions) or emitted interference in terms of ELF (extremely low frequency) electric or magnetic fields or RF (radio frequency) electromagnetic fields such as from diathermy units or RF surgery devices. Emissions might adversely affect other devices or expose persons to levels even above existing safety levels /50/, /52/ and might cause adverse health effects.

Impact on the environment could be caused by:

- ***powerful devices*** (e.g. X-ray devices) causing adverse changes and fluctuations of the mains voltage with the risk of consequential damage of electronic components or their malfunction. As an example, the X-ray spectrum and consequently X-ray image quality and patient exposure critically depend on the X-ray anode voltage and in turn on mains voltage. To limit disturbances, conducted interference is limited.

Remark: Voltage fluctuations of the power supply system are also due to voltage drops caused by electric currents flowing across the internal resistance of the power supply system. In the course of periodic inspection, the internal resistance of the power supply system is measured between live conductor and neutral conductor.

- ***electrostatic discharges*** may emerge from contacting and subsequent separation of materials. This happens for example when moving devices with poor conducting wheels over electrically insulating floor coverings. The amount of charging depends on the velocity of the devices and their electric capacity. To avoid charging, manufacturers are requested to mount electric conducting wheels to reduce a device's ground resistance to lower than 10 kΩ /55/, /60/.
- ***magnetic emissions*** from medical devices can affect functioning of other medical devices or active medical implants (e. g. cardiac pacemakers or cardioverters). Extremely high magnetic fields (e. g. magnetic resonance imagers) could require particular precaution. As an example, in the area up to several metres from MRI devices magnetostatic inductions above 0.5 mT could activate reed relays of cardiac pacemakers and cause changing their operation to the fixed-frequency asynchronous interference mode. Mains frequency magnetic fields are caused mainly by electric power cables, transformers and motors. At the enclosure of electric appliances magnetic fields could reach several mT (averaged over 100-cm^2 sensing area). Although decreasing with the square of distance such fields could interfere with neighbouring medical devices. In spite of inherent mains frequency notch filters biosignal amplifiers may remain vulnerable to harmonics. A market survey demonstrated that magnetic emissions of electric appliances even of the same kind may vary by up to two orders of magnitude. Minimizing them is no design criterion as yet /47/ neither for household appliances nor for medical devices.
- ***electromagnetic fields*** intentionally generated by medical devices (e. g. diathermia, RF-surgery, hyperthermia) may cause strong stray fields outside the region of interest leading to interference with other devices (e. g. nerve and muscle stimulators) and even overexposure of nearby persons and of vulnerable regions of the patient outside the target area (e. g. brain, eyes, testes, gonads) /46/. Electromagnetic fields with frequencies up to several hundred kHz can also be generated by sparks such as produced by DC motors, for example of centrifuges.
- ***emitted heat*** may impair the functioning of other nearby devices by thermal offset, drifts or straining their cooling.
- ***emitted radiation*** (laser radiation, UV-radiation, X-rays or gamma-radiation) might pose risks to persons even up to a large distance.
- ***chemicals*** may be emitted from materials or substances due to diffusion or leakage [e. g. Cr, Cl or hydrocyanic acid (HCN)] from gas lasers, anaesthetic gases might leak from devices or from a patient's breathing mask]. Plastic may emit adverse substances, for example softeners (e. g. phthalates). Chemical pollutants may be set free also in single fault condition or after disposal (e. g. mercury-vapour pumping lamps of lasers or for light therapy, electronic components or batteries). Manufac-

turers must minimize such risks by environmentally conscious design and defining installation and/or air conditioning of rooms (see Chap. 7).

6.2.2 Fire and explosion protection

Fire and explosion have the same cause, namely ignition (oxidation) of substances. Whether ignition develops into fire and fire develops into explosion depends on whether burning can continue, and if so, on its rapidity. Risk analysis has to include fire hazards. For instance, flammable covers are unacceptable for wheel chairs since smoking by the patient (ignition source) or an accompanying person cannot be excluded. A warning in the instructions for use (as shown in the text box) is insufficient for risk control because of the manufacturer's obligation to prefer constructive safety over warnings wherever possible and reasonable.

> **Warning:** The cover of our classical wheelchair is made of easily flammable material. Therefore, smoking is prohibited.

Medical devices must be designed so as to prevent them from becoming a source of fire (§ 11.2 EN 60601-1). This is possible by different approaches such as avoiding flammable material, dangerous ignition sources (e. g. by limiting electric energies), preventing propagation (and limiting available oxygen) by using inflammable enclosures, and minimizing operation time (Figure 6-4). Elevated risks must be expected at devices of long-term uninterrupted operation (e. g. devices remaining in standby also in uncontrolled periods, overnight and/or over weekends).

Explosion

Explosions differ from fire by their dynamic progress. Explosions are characterized by rapid oxidation and the following increase in temperature and pressure. However, both, fire and explosions are only possible if three conditions are met simultaneously and

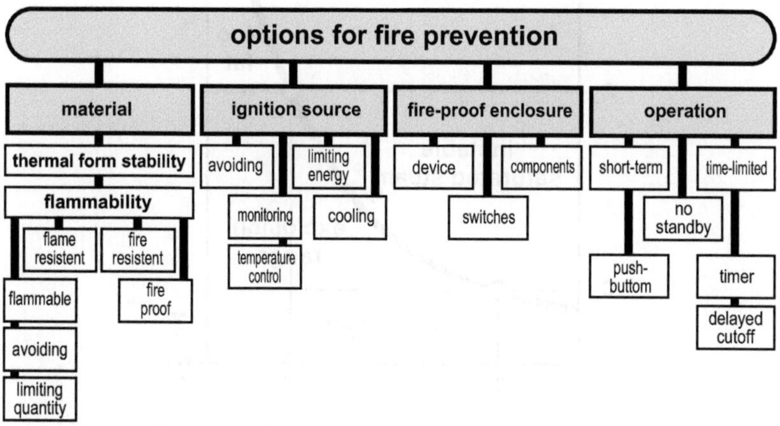

Figure 6-4: Options for fire prevention

sufficiently which are a flammable substance, oxygen and an ignition source (Figure 6-5). Therefore, fire can already be prevented, if only one of these conditions is not fulfilled. This is the reason, why highly explosive applications such as RF surgery are possible even in areas with high explosion hazard (e.g. patient's gastrointestinal tract). If inert gas (e.g. argon, nitrogen) is blown into the surgical area oxygen is removed and one of the three requirements is not met. Consequently, explosion is made impossible.

Since an appropriate mixture of flammable substance and oxygen is required, it is impossible to directly ignite solid matter and liquids. Before this can occur, it is indispensable to increase the oxygen content, for example by either pulverizing material or evaporating liquids. Since ignition is prevented by too low concentration c_u (too little flammable substance) as well as by too high concentration c_o of a flammable gas (too little oxygen), even mixtures of a flammable gas and oxygen can be ignited only in a limited concentration range, the "explosion range" (Figure 6-6).

In medical locations explosive mixtures of flammable gases (e.g. anaesthetic gases, steam of disinfectants or endogenous gases) may be encountered with air or with pure oxygen or laughing gas (N_2O). Presently used anaesthetic substances in their usually

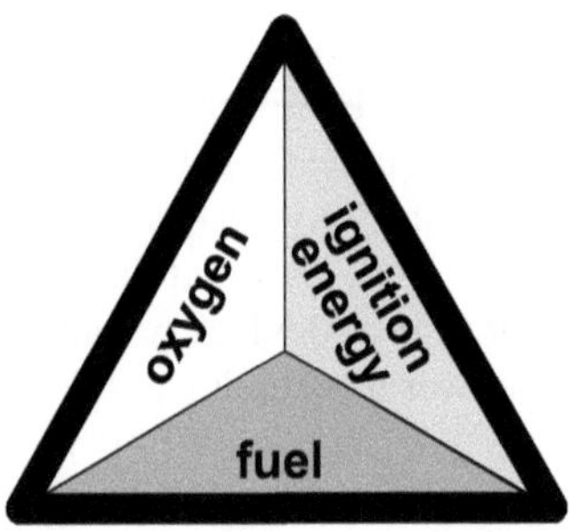

Figure 6-5: Three conditions for fire and explosions

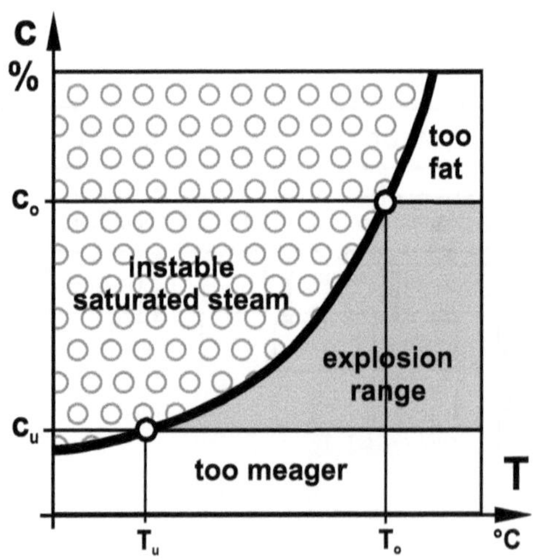

Figure 6-6: Explosion range of a mixture of a flammable gas and oxygen in dependence of concentration c and temperature T.

applied concentrations are hardly flammable. However, energies required to ignite detergents or disinfectants are much lower than the electric energy released from a person's electrostatic discharges which can reach up to about 20 mJ. Hot surfaces such as those within electronic devices already in normal condition can also become dangerous ignition sources. Surface temperatures above 160°C can already be sufficient to ignite the steam from disinfectants (Figure 6-7).

Remark: Surface temperatures of 40 W bulbs (155°C) are below, but of 100 W bulbs (260°C) and low-voltage halogen lamps (270°C) are well above this ignition temperature.

Manufacturers must be aware that components with sufficient high temperatures are frequently found in devices already under normal conditions (e.g. power resistors, power amplifiers, transformers) and – of course even more dangerous – in single fault condition.

Remark: Components of medical device's are permitted to reach temperatures of up to 180°C (NC) and even 260°C (SFC).

Explosive regions

In medical locations explosion hazards occur in particular at places where flammable gases (e.g. anaesthetic gases) or liquids (e.g. disinfectants) may be used such as in operating theatres, intensive care units, anaesthetic recovery rooms, surgical ambulances and delivery rooms. However, also our body generates flammable gases making the gastrointestinal tract and the upper airways potential explosive areas.

Figure 6-7: Minimal ignition energy $W_{Z,min}$ and ignition temperature T_Z of detergents and disinfectants. Open circles … non-alcoholic substance, full circle … endogenous gases, triangles … alcoholic substances, Ä … ether, C_2H_2 … acetylene, H_2 … hydrogen

For safety considerations two different explosive zones are distinguished:

Zone M (*medical environment*): It comprises regions where for short time periods explosive mixtures of flammable gases (detergents, disinfectants, anaesthetic gases) with air may exist. This must be assumed in particular in operating theatres above treated skin during evaporation of disinfectants, around a patient's respiratory mask and around breakable parts enclosing zone G regions. Since flammable gases are heavier than air they can accumulate below the operating table and increase explosion hazards there (Figure 6-8). This is of particular importance if devices are on the floor which might release ignition energy such as switching sparks of footswitches.

Explosion hazard can be banned by diluting concentrations to uncritical levels with sufficient air exchange rates. This can be assumed at exchange rates above 15-fold (in case of external air aeration) or 60-fold (in case of recirculatory air aeration). Explosion may also occur by ignition of endogenous gases that are accumulated within the gastrointestinal tract and released to the upper airways.

Zone G (*enclosed medical gas system*): It comprises completely or partly enclosed cavities or regions where explosive mixtures of flammable gases with pure oxygen might continuously or temporarily be produced, transferred or applied. This must be assumed in patient's upper respiratory tract (upper airway including the lung). During application of devices producing high ignition energies (e. g. RF surgery, laser surgery) in explosive zones oxygen must be removed around the applied part by protective gases (e. g. argon, nitrogen).

Patient burned at OP table

The Hague: A 69-year-old patient was locally anaesthetized and strapped to the operating table for routine surgery when a flash rushed from the anaesthetic device. The fire was speeded up by leaked oxygen. Burning progressed so quickly that the nine present persons were not able to extinguish the flames quickly enough. The women burned alive.

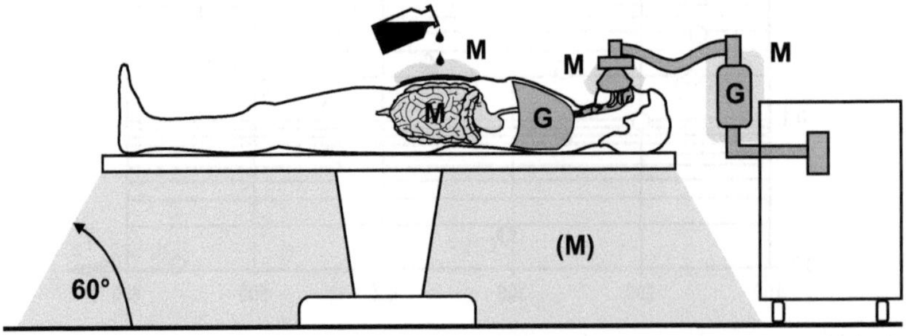

Figure 6-8: Explosive zones M and G in the operating theatre. Zone M below the operation table can be avoided by sufficient air exchange.

Oxygen

Oxygen is vitally important for us. However, it can also be very dangerous. This is because pure oxygen mixed with flammable gases critically enhances explosion risk. The reason is that oxygen makes fire burn easier, quicker and hotter. Since oxygen is heavier than air, if leaking, it accumulates at the bottom of a device or on the floor. Oxygen cannot be smelt. Therefore, we cannot perceive dangerous oxygen concentrations.

The oxygen concentration of air is 21%. Already a few percent more increase fire danger considerably by critically reducing ignition energy and the required ignition temperature. This makes otherwise harmless ignition sources (e. g. cigarette glow, light bulbs) dangerous and otherwise harmless material (e. g. clothing or PVC-insulation of electric conductors) easily flammable. Fat, lubricants but also beauty creams or ointments may react with oxygen even explosively. Therefore, only such oxygen bottles, fittings and pressure reducers should be used that are intended for oxygen. They have to be oil- and fat-free.

! Oxygen must not be brought into contact with fatty substances

Safety standards define air with an increased oxygen content of 2% (from 21% to 23%) as oxygen-enriched atmosphere. Medical device standards consider regions with oxygen content increased by 4% (from 21% to 25%) as oxygen-enriched environments (§ 3.75 EN 60601-1) which require particular preventive design to reduce the otherwise unacceptably high fire danger (§ 11.2.2.1 EN 60601-1).

Preventive design aims at avoiding potential ignition sources (e. g. switching sparks or hot components), encapsulating them, minimizing their energy to prevent from dan-

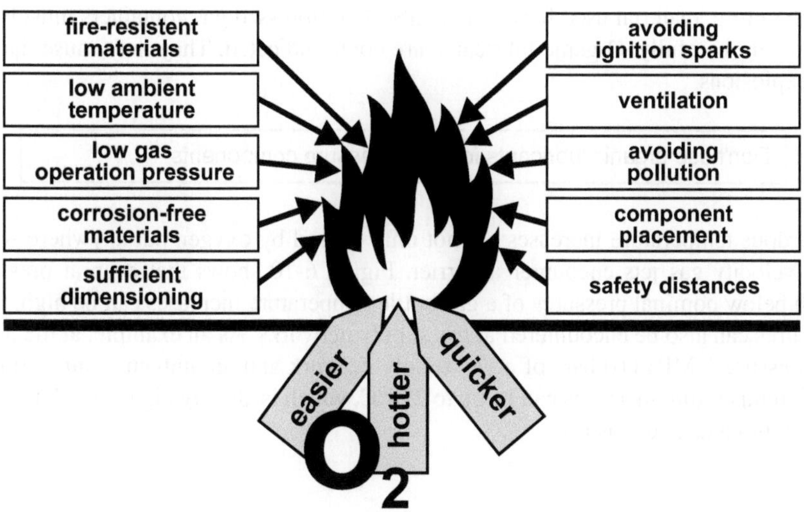

Figure 6-9: Design options reducing risk from oxygen

gerous consequences of an explosion (Figure 6-9). Device enclosures must not be flammable. Devices with the risk of oxygen leakage (and accumulation) require sufficient ventilation. Electric plug connections (that may be associated with disconnection sparks) must keep a safety distance to oxygen connectors of at least 20 cm.

If oxygen is under high pressure like in gas bottles (e.g. ventilators, anaesthetic devices) ignition hazard is considerably increased. If the valve is opened too quickly, gas flashes to the pressure reducer with high speed and compresses the gas in between too quickly to allow heat exchange with the environment (adiabatic compression). As a consequence, temperature (in degree Kelvin °K) increases to high end values T_{End} (Figure 6-10) according to following equation

$$T_{end} = T_0 \cdot \left(\frac{p_{Fl}}{p_0}\right)^{0,29} \quad \text{with T in °K}$$

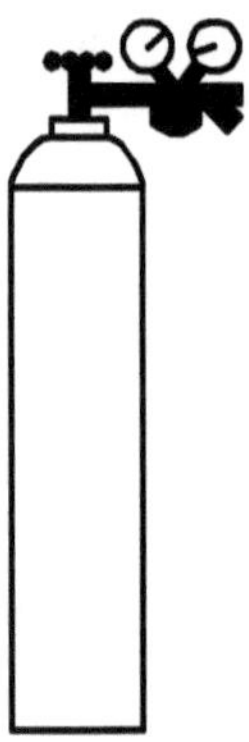

At room temperature T_0 of 20°C the rapid opening of a valve of an oxygen bottle with the pressure p_{Fl} = 200 MPa (200 bar) leads to a temperature increase up to 1,089°C. Not even many metals can withstand such a high temperature.

As examples, self-ignition of metallic material starts already at about 900°C /22/. In air, the ignition temperature of iron amounts to about 1,050°C; lubricants can be ignited already above 250°C, PVC above 380°C, PP at about 150–160°C or PE at about 180–200°C.

This demonstrates that even metals can become dangerous ignition sources or can even burn themselves (oxidize) with the danger of triggering reaction chains. Rapid opening and the following adiabatic compression may ignite a small non-metallic component causing excessive heating which in turn could ignite the metallic housing of the valve which frees the way to the compressed gas. As a consequence, the gas bottle may fly away like a rocket and could lead to mechanical damage. In addition, released oxygen would critically increase fire hazard.

Therefore, material used for components of compressed gas containers must be selected very carefully. Organic lubricants are contraindicated. They may cause dangerous explosions.

! Don't use organic lubricants for high-pressure components

Hazardous temperature increases are not only caused by oxygen but anywhere where high velocity gas jets encounter a barrier. Figure 6-10 shows that even at pressures much below nominal pressures of a gas bottle temperature increases can be high. High pressures can also be encountered at gas supply networks. As an example, at the nominal pressure 1 MPa (10 bar) of a gas supply network and an ambient temperature of 20°C, temperature increases can be up to 298°C, which is already high enough to ignite in air lubricants and plastics.

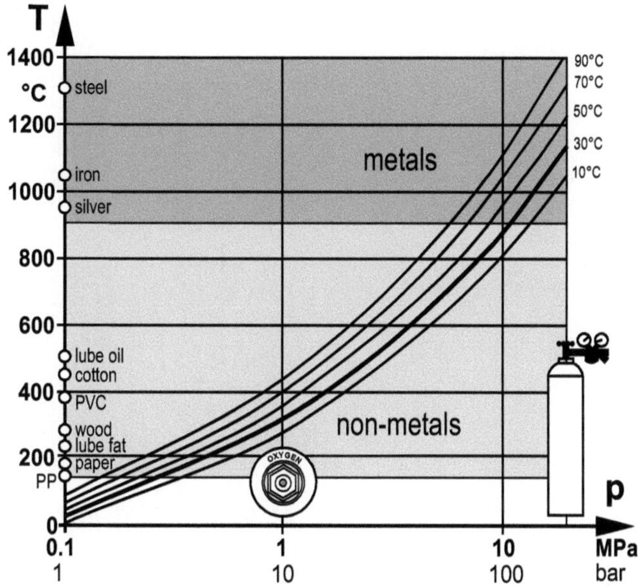

Figure 6-10: Temperature T caused by quick opening of an oxygen bottle valve (adiabatic compression) in dependence on bottle pressure p for different environment temperatures with ranges of self-ignition of non-metallic (grey) and metallic material (dark grey). Parameters … ambient temperature in °C

7 Ecological safety

Manufacturers must design medical devices in an environmentally conscious manner to minimize adverse effects on the environment and ecology /31/. This requires recognizing environment-relevant aspects under normal and single fault condition, for example in the case of fire. Such aspects do not only comprise the release of adverse substances in the form of gases, vapour or dust. They include also the conscious use of natural resources (energy, raw material), generation of noise, vibration or electromagnetic fields, and safe disposal after the end of service life. Therefore, environmental protection influences selection of used material, design, construction and manufacturing, decisions relating to restricted/unrestricted service life time, reuse or single-use, packaging, installation, use and maintenance of medical devices including recycling and disposal (EN 60601-1-9 /31/). The objective is to minimize release of critical substances which could be adverse to health because of microbial contamination, toxicity or environmental incompatibility. One option for environmental protection is shifting tasks from hardware to software.

According to the European Directive WEEE 2002/96/EG /9/ manufacturers have to take care of ecological disposal and/or enable reuse or recycling of waste equipment. Refurbishment and recycling make it possible to reutilize valuable raw materials in particular if design has considered this aspect in a timely manner and easily allows separation of components and selective reuse and/or disposal (Figure 7-1).

Manufacturers must be aware that electronic devices contain a manifold of partly highly toxic substances which might be directly released into the air, at contact areas diffuse into human body tissue or may be released into the ground when deposited. Critical components are numerous:

- plastic enclosures and insulation may contain over 50% dangerous additives, e. g. flame-retardant halogens, softening phthalates, metallic soaps or stabilizing lead salts;
- heating devices may contain asbestos to insulate heat;
- solders may contain lead;
- electronic components frequently contain problematic substances, e. g. electronic boards (flame-retardants), capacitors (electrolytes such as polychlorinated biphenyles (PCB), tantaloxide, manganoxide), LEDs and displays (gallium arsenide), relays (mercury);
- electronic components may contain even dangerous substances, e. g. batteries (mercury, cadmium, zinc, lithium), transformers (PCB-impregnated paper) etc.

Therefore, in general similar to other electronic devices electromedical devices have to be considered hazardous waste, which has to be disposed of according to the relevant

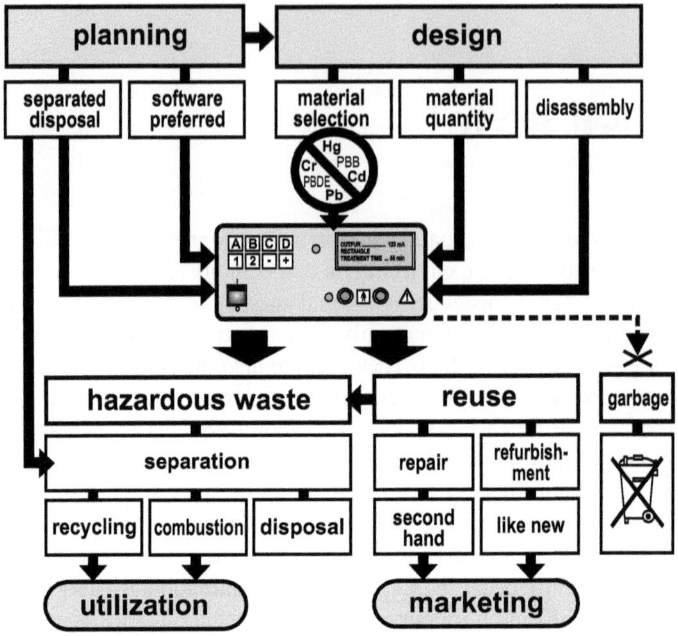

Figure 7-1: Aspects of environmentally conscious device design

local regulations. A warning should be given not to dispose of it as consumer waste. This has to be indicated in the instructions for use and marked on the devices.

The European directive on the restriction of use of certain hazardous substances in electric and electronic equipment (RoHS 2002/95/EG) /10/ requests that manufacturers of electronic devices avoid substances hazardous to health or environment through adequate design. Some substances are banned such as mercury, cadmium, hexavalent chrome, polybrominated biphenyls (PBB) and polybrominated diphenyl ethers (PBDE). Exempted are such applications where for technical or scientific reasons substitution is impossible or would lead to even worse consequences. As an example, while mercury thermometers are banned, mercury-containing fluorescent tubes or energy-saving lamps as well as lead-glass X-ray tubes are permitted.

8 Electric safety

Electricity is similar to truth: Everybody talks about it, but nobody really knows it. Today, we know that electricity is one of the very few fundamental phenomena of nature enabling the very existence of material and functioning of life. Every substance is composed of atoms consisting of a positively charged nucleus surrounded by a cloud of negatively charged electrons. We know the effects of electricity and as far as possible succeeded in making its forces useable, but the question of what its final nature is, remains open. We know that electric phenomena are caused by "electric charges," and that there are two different kinds if them. Since the 18th century (without any intention of making value judgements), the two kinds of charges have been called "positive" and "negative". We also know that different materials, although appearing neutral, contain balanced numbers of both kinds of electric charges, although with different density. This explains why we may experience annoying discharges when we get out of plastic seats (see Chap. 6.1.3). The reason is, that at the contact area of different materials electric charges are exchanged to balance different charge densities – with the consequence that now both materials loose neutrality and become charged. If materials are separated again and this happens quickly enough and/or the mobility of charges is low, the migrated charges cannot flow back to their initial material and the surplus may discharge by an electric spark to earthed parts (e. g. the doorknob).

In ancient Greece there were, of course, no plastic chairs, but people already knew that amber could be brought into a different "status" just by rubbing with wool. This explains the origin of the term "electricity." It is derived from the Greek word for amber, "electron" (ελεκτρον).

Remark: Electrostatic charging by friction is common in daily life, but it can lead to severe safety problems in medical locations. The reason is that in an operating theatre discharging energy of a single person is far more than sufficient to cause explosions of flammable mixtures. This is the reason why antistatic measures are required both for installations and medical devices (see Chap. 6.1.3)

Although we do not know the final nature of electric charges we have very detailed knowledge of how they behave. They are quite similar to other things in life, where opposites attract and similarities repel each other. In electric technology the unified strength of these interactions is called electric field strength and the electric field described graphically by lines of the potentially acting electric field strength (Figure 8-1).

! Opposite charges attract each other, similar charges repel each other

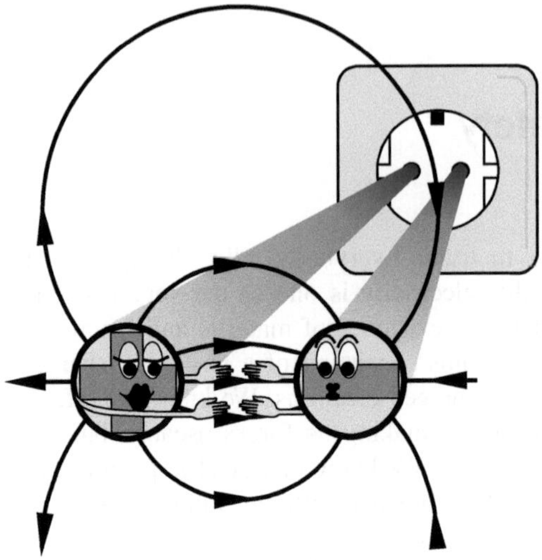

Figure 8-1: Electric field between two differently charged electric particles

The principle of attraction and repulsion governs all electric technology. However, once two different charges are combined, something perplexing happens. In a surprising magic trick of nature, both charges suddenly disappear from the (electric) scene as if they had never existed: A pair of opposite electric charges behaves as if they were electrically neutral.

However, separating opposite electric charges is laborious. It requires work (which must be done, for example, in power plants) and is not finished with mere separation. In addition, premature reunion of charges must be prevented by electrically insulating them from each other. The reason is, that separated charges try to reunite, the more they are separated the stronger the attempt to reunite. This status can be characterized by an electric voltage which increases with the number of separated charges.

Especially in regard to the effects of electric voltage and electric current there is widespread confusion. It is important to emphasize that the mere existence of an electric voltage does not necessarily already cause harm. It solely describes the hazard, the precondition for it. The effect itself is caused by other parameters.

! Electric voltage is (just) a precondition for but not the causation of harm

The fundamental principles of electric currents can be explained by their similarity with water flow. Therefore, electric effects can be illustrated by taking advantage of an electric-hydraulic analogy considering the terms summarized in Table 8-1 as similar to each other.

A battery or a mains socket outlet has two poles at which unipolar charges are accumulated. This can be compared to a dam with retained water behind it. The thickness of the dam is dependent on the height difference from water level to ground. Likewise, the strength (thickness) of an electric insulation depends on the electric voltage. Harm is caused only if the dam breaks and water floods the valley. This is similar to electric-

Table 8-1: Analogy between electric and hydraulic parameters

Electric parameter	Hydraulic parameter
electric charge	water drop
electric potential	water height
electric voltage	water pressure (height difference)
electric current	water flow
electric resistance	flow resistance
electric insulation	water tube

ity. An insulation failure would give way to rapid reunion of "retained" charges and allow a high electric current flow (short-circuit current) with potentially hazardous consequences. The resulting effect depends on the number of charges involved. However, in contrast to batteries at mains socket outlets this number is almost infinite. To act as a charge limiter (to limit short-circuit currents) in electric installations overcurrent circuit breakers are of ultimate importance.

If you wet flowers with a garden hose, you take advantage of the same principles applied to electric currents. Water is provided at high pressure. Reduction of the strength of the effect to softly water flowers rather than bending them depends on the amount of water and its velocity (the strength of the water jet). Likewise, the hazard of electricity depends on the amount of charges and their velocity (the amplitude of the electric current) flowing across our body.

! Effects are caused by electric currents

The strength of a water jet depends on the water pressure and, hence, the flow resistance, which means, how far the water hose has been opened and flow resistance reduced. In a similar way electric current and electric voltage depend on each other with the electric resistance being the proportionality factor. This finding is summarized by the fundamental law of electro technology, Ohm's law, namely electric current equals voltage divided by resistance.

! current = voltage / resistance

Although simple, this relation has far-reaching consequences. First, it means that electric voltages would require an infinite high resistance to prevent currents from flowing. However, such an ideal infinite resistance does not exist. It follows, that as a principle, electric (leakage) currents cannot be avoided and are encountered everywhere in daily life where electric voltages exist.

! Electric (leaking-) currents are omnipresent

A second consequence of Ohm's law is that at the same voltage hazard increases with reducing resistance. Therefore, if the electric resistance of our insulating dry skin is reduced because it has become wet during bathing or cooking, electric hazard has also increased. Consequently, danger is highest in medical locations where the protecting skin has been injured or the body opened during a surgical operation. This needs particular consideration when designing medical devices.

In contrast, electric safety increases with increasing insulating properties of the skin such as by use of fatty creams or by wearing insulating gloves for example during homework or surgery.

However, Ohm's law can also be written in a different way, namely that voltage is caused by currents flowing across resistances.

> **!** voltage = current x resistance

At first glance it might not be obvious that important conclusions follow from this simple relation. However, since there are no ideal insulators – under daily life conditions – there are also no ideal electric conductors either where electric currents would not encounter a resistance. From this, it follows that every current unavoidably causes an electric potential difference (voltage) and new hazard potentials (see Sect. 8.5.1). It was already stated that electric currents are omnipresent. Consequently, this applies also to (unintended) electric voltages. These can be high enough to require additional safety precautions (e. g. potential equalization) both in daily life as well as in medical locations (see Sect. 8.2.2).

> **!** Electric potential differences (voltages) are omnipresent

From time to time newspapers wonder with curiosity that even within their homes people might be killed by a lightning strike in spite of an intact lightning protective system (see text box). However, this can be explained by the consequence of Ohm's law and the indirect effect of an unintended potential increase. Lightning currents could reach enormous amplitudes up to 500,000 A. If such a high current hits the lightning arrester and flows to ground, even with the low grounding resistance of the lightning protective system (and the electric installation) dangerous situations could arise. As an example, if a lightning current of 500,000 A flows across a grounding resistance of 5Ω due to the voltage drop the potential of its upper end (the "earth potential" of the electric installation) now increases to a high potential. It now exhibits a potential difference to ground of 2,500,000 V. The consequence is that this enormously increased potential to ground is now shared by all protectively earthed parts (e. g. enclosures of appliances) within the whole house, while other parts like water pipes or heaters remain at their initial ground potential. As a consequence, lethal potential differences (voltages) can appear within homes. In the given example, the voltage would be even high enough to exceed the withstand voltage of air and cause indirect strikes of lightning to appear inside homes. They could be lethal and even cause fire. To avoid such dangerous situations, it is mandatory that ***all*** parts within a house exhibiting connections to ground (e. g. water supply system, heating system, electric installation) be connected to each other to equalize their electric potential. This potential equalization assures that in case of a lightning strike all these parts are elevated to the same (high) potential, but dangerous potential differences (touch voltages) are avoided.

Lethal lightning inside home

Innsbruck: A housewife was found dead in her kitchen. The initial assumption of an electric accident proved not to be true. Investigations showed that the house had been hit by a lightning strike. This indirectly caused a lethal electric shock within the kitchen because of selective potential increase due to the voltage drop at the earth resistance caused by the lightning current. This could have been avoided if the electric installation had met the standard.

There is still one seeming contradiction to be resolved. If somebody carries a car battery and touches the positive pole, nothing happens, and no current flows across the body. However, as soon as the battery is connected to the car and the same pole is touched again, one could experience an electric shock due to current flow. This observation is of considerable importance. The explanation is that currents can only flow, if the electric circuit is closed and allows separated charges to recombine with opposite charges at the counterpole.

! Electric currents can flow in closed circuits only

It is no contradiction to the above statement that electric currents in fact did flow across the body when touching one pole of the car-mounted battery. The reason is that the negative pole of the car battery is connected to the coachwork, and successively to ground. Because for safety considerations persons are generally considered to be grounded, when touching the plus pole, current is flowing from the plus pole via the person to ground and via the coachwork back to the minus pole – which closes the circuit.

! For safety considerations persons are considered to be grounded

The seemingly trivial requirement that electric currents can flow in closed circuits only has two important aspects: It offers an opportunity and could be a danger.

1. the **opportunity** is that it can be used for an important safety strategy. Namely, if all poles of an electric current source are insulated from earth, a person (who is considered to be earthed) remains protected when touching one of the poles, because the circuit still remains open. This is not just a theoretical sophistry. In fact, this safety option is chosen for power supply of operating theatres where a safety insulation transformer is used to generate an earth-free power supply system (IT-net). This allows the killing of two birds with one stone. On the one hand, in addition to the safety measures of devices, electric shock protection is increased. On the other hand, in single fault condition short-circuits and activation of circuit

breakers with subsequent blackout of all attached devices can be avoided, and undisturbed continuation of work and operation of life-supporting devices assured (see Sect. 6.1.2).

Remark: In general, in daily life in homes and in medical locations the electric installation for several technical reasons is earthed (with one pole connected to ground). This has the disadvantage that persons accidentally touching live parts are in danger. For economic reasons earth-free voltages are provided only in exceptional cases such as in bathrooms where socket outlets for shavers are equipped with an integrated insulation transformer.

2. the ***danger*** of the fact that electric currents can flow in closed circuits only stems from the consequence that electric currents use ***all*** existing possibilities to flow to the counterpole, not just the initially intended one. Ohm's law demonstrated that it is not possible to prevent currents from flowing even across dedicated electric insulation, and because ideal insulation does not exist, there are always many pathways, currents could choose. Similar to a "rush hour," when traffic uses even small side streets to move forward, also electric currents use any route to flow back to the counterpole. Ohm's law and the electric resistances just decide upon the partition of currents. Therefore, if touching a device or an applied part, of course, electric (leakage) currents flow also across persons, and their amplitude is the greater the lower the (body) resistance is.

! Electric currents use all possible pathways to the counterpole

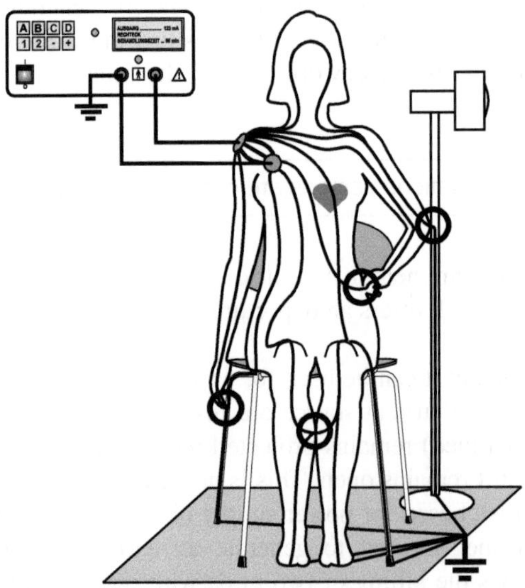

Figure 8-2: Unintended bypasses during nerve stimulation with a device of an earthed patient circuit and indicated points of enhanced current densities at small-area contacts (circles)

This principle is particularly important for patients who are intentionally connected to a current source (e. g. to a muscle stimulator). Since currents are not restricted to flow the intended way only, particular care must be taken to avoid hazards from currents flowing along bypasses offered by accidental contacts to other parts of the body (e. g. by extremities) or to electric conducting parts (e. g. operating table, chairs or supports). Dangerous bypass currents may also occur if at the same time several medical devices are applied to the patient offering several new possibilities for flow of bypass currents (e. g. anaesthetic device, ECG monitor, blood pressure monitor, blood gas monitor, suction device, RF surgery device). The situation would become extremely complicated if patient current circuits were earthed and currents could flow back across any connection to earth (Figure 8-2).

For these reasons electric circuits intended to include the patient (e. g. nerve and muscle stimulators, biosignal recorders) must be insulated from earth and have applied parts type BF or CF (see Sect. 8.3.2).

8.1 Biological aspects

8.1.1 Body resistance

We know that electric voltages are not directly causing effects but are just determining the hazard potential. If somebody touches live parts, the electric (body) resistance determines whether electric currents flowing across the body can reach dangerous values or not.

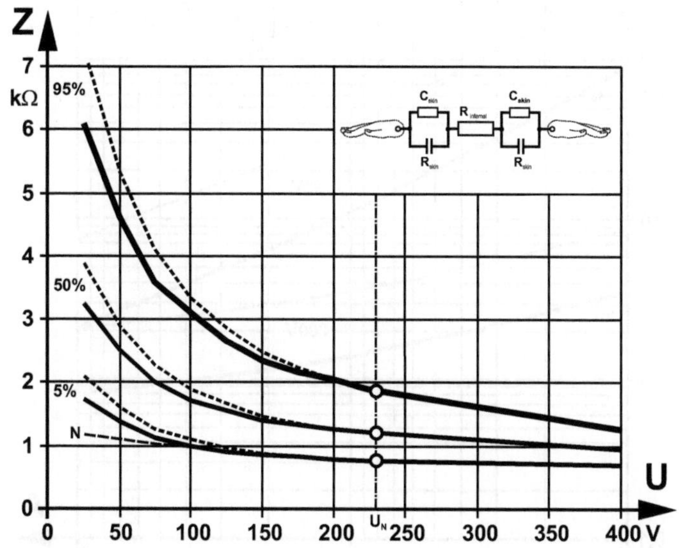

Figure 8-3: Whole-body impedance Z for hand – hand current pathway in dependence of voltage amplitude U for dry skin. Percentages indicate the partition of persons with impedances up to the value. (full line … alternating voltage; broken line … direct voltage), N … wet skin (derived from EN 60479-1 /26/)

At the beginning of the 20th century first measurements of the electric body resistance ware made on dead bodies, but investigations on volunteers were necessary to clarify whether and to what extent such results could be extrapolated to living humans /1/. Today we know that apart from the internal electric resistance it is the resistance of the skin that considerably contributes to electric shock protection. The different layers, the dry horny skin, followed by the epidermis, the dermis and subcutaneous insulating fat layer – from the electrotechnical point of view – generate a capacitor which is a frequency-dependent component of decreasing impedance with increasing frequency. At low direct voltages the whole-body impedance is abut 20% higher than at alternating voltages.

It was observed that there is also another reason why the whole-body resistance decreases with increasing voltage. The reason is that insulating skin layers cannot stand high voltages. Therefore, with increasing voltage spots with electric breakdowns accumulate and skin resistance decreases. Since the insulation ability of the horny skin layer decreases with humidity, wet skin exhibits a lower impedance than dry skin (Figure 8-3).

Skin impedance critically depends also on the size of the contact area (Figure 8-4). If the contact area is reduced from palm (about 100 cm²) to finger contact (about 1 cm²) the impedance increases 4-fold (at 200 V) and at low voltages (20 V) even to 40-fold.

For safety considerations the whole-body impedance is defined to be 2 kΩ at mains frequency (50/60 Hz). It is assumed that skin impedances at both contact sites contribute 50% to this value (Figure 8-5).

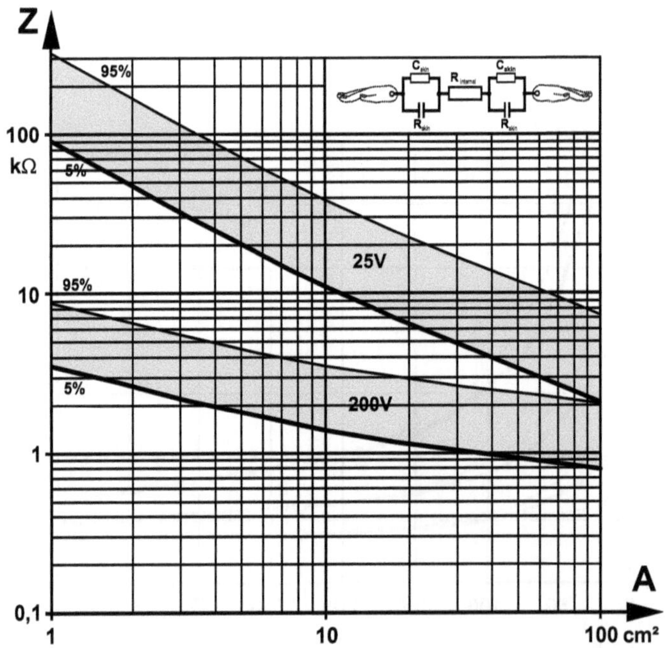

Figure 8-4: Whole-body impedance Z for hand – hand current pathways in dependence on contact area A for dry skin and two different voltage levels. Percentages indicate the partition of persons with impedances up to the value (derived from EN 60479-1 /26/).

Whole-body impedance ≈ 2 kΩ = 2x skin impedance + internal body resistance

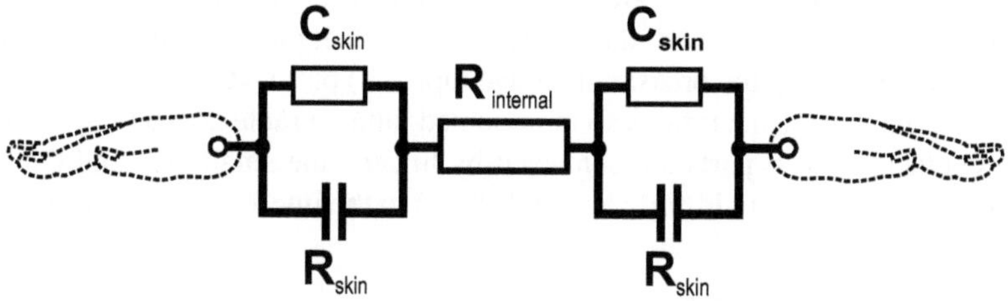

Figure 8-5: Equivalent circuit diagram of the whole-body impedance

In the low-frequency range the ***internal*** body impedance is not dependent on frequency. Since an electric resistance increases with decreasing cross-sectional area available for current flow, contributions to the internal body resistance are highest at joints with almost no muscle tissue and poor conducting bones and tendons, in particular at the wrist and ankle. Therefore, extremities with their joints dominate the internal body resistance. For safety considerations it can be characterized by a simplified star-shaped equivalent circuit diagram where the thorax resistance is neglected and extremities are represented by four resistances of 500 Ω each (Figure 8-6).

The degree of electric shock protection by the ***internal*** body resistance is dependent on the kind of contact, voltage height and electric current pathway across the body. For

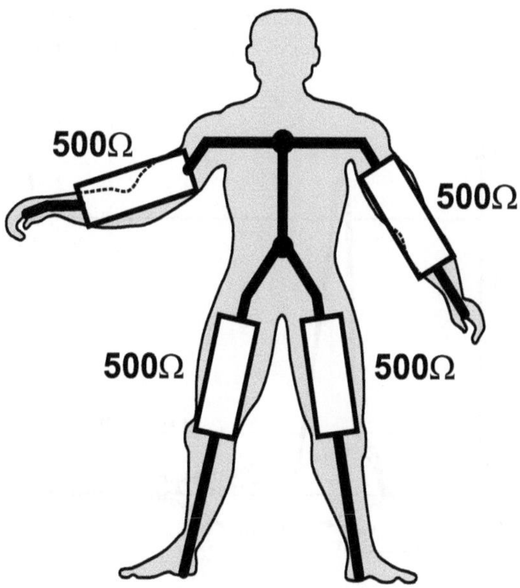

Figure 8-6: Simplified equivalent circuit diagram of the internal body resistance

hand – hand contact the internal body resistance amounts to 1,000 Ω (Figure 8-6). If the current flows from one hand to both feet, duplication of current pathways along both legs reduces the resistance to 750 Ω. Even more unfavourable is touching live parts with both hands and with both feet on ground. In that case the internal body resistance is reduced to 500 Ω. The internal body resistance can be even lower such as 250 Ω in the case of contacting a live part with the breast with both feet on the ground or about 10 Ω in the case of contacting the breast with a live (applied) part and the back grounded by the operating table. Considerations so far assumed unfavourable contacts by sufficiently large contact areas. If parts are contacted by fingers, the small area and the higher resistance of the finger would add another 1,000 Ω to the internal body resistance.

8.1.2 Cellular excitation

It was the year 1780. Spooky silence filled the room and a flickering candle produced dancing shadows as Luigi Galvani became startled. A dead frog's thigh seemed to have regained life and performed a convulsive movement when an electric spark jumped over to it. Galvani was convinced he had discovered the basis of life in terms of "animalistic electricity." Today, we know that his assumption is not true, but it is fascinating enough to have evidence that our cells are like small batteries exhibiting a potential difference between the interior and exterior (membrane potential) of about –90 mV. The effect that had horrified Galvani can be explained by the fact that physiological processes are fundamentally based on electricity, and that nerve and muscle cells can be excited by electric currents which change the cellular membrane potential. While small currents cause only unspectacular changes, a "local response," the situation changes dramatically if current amplitudes exceed a characteristic value, the excitation threshold. In this case a chain reaction is triggered which finally leads to a characteristic impulse-like change of the membrane potential, an "action potential" (Figure 8-7).

If nerve cells are involved such a "nerve impulse" does not remain at the site of excitation but is transmitted along a nerve fibre. This allows rapid transport of mes-

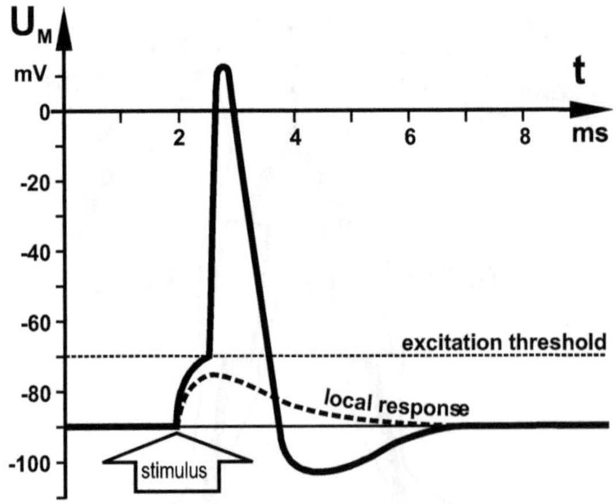

Figure 8-7: Time course of the electric potential at the membrane of a nerve cell following subthreshold stimulation (local response) and above-threshold stimulation (nerve impulse or "action potential")

sages from periphery to brain and vice versa with speeds higher than that of a formula 1 racing car, up to about 430 km/h. Once the excitation threshold has been exceeded the shape and amplitude of an action potential does not further depend on the strength of the stimulus. This is called the "Law of all or nothing." Strong stimuli do not cause other action potentials but only speed up their generation and hence their repetition frequency.

Cells do not react to any stimulus. Since cellular excitation takes time, it may stay absent if stimuli end too soon. Even increasing the stimulus amplitude can only partly compensate for decreasing its duration. Therefore, the duration of a stimulus must exceed a certain minimum to excite a cell.

Our optical, acoustic, tactile and thermal sensors continuously flood the brain with nerve impulses. This information flow amounts to about 10 Mbit/s. If we had to consciously process all these data we would have a severe problem. This would not only overload our perception capacity but also it would be very difficult to select what is important and what not. In fact, we are only able to consciously perceive a small amount of incoming information, only about 17 bit/s. Obviously, this continuous reduction of information must be enormously efficient and still allow separating important from unimportant messages. One principle that assists us in achieving this filtering performance is to suppress constant stimuli. This explains why after a while we are no longer aware of the constant noise or we no longer smell the smelly air in a pub. This is achieved because nerve cells adapt to constant or slowly changing stimuli and require a minimum rate of change for excitation.

From the above, it follows that for excitation of nerve and muscle cells a stimulus must sufficiently meet three conditions: It must exhibit a sufficient strength, duration and change (Figure 8-8). Consequently, direct currents as well as high-frequency currents cannot cause cellular stimulation: direct currents due to lack of change and high-frequency currents because of the too short duration of their excitatory half wave (Figure 8-9).

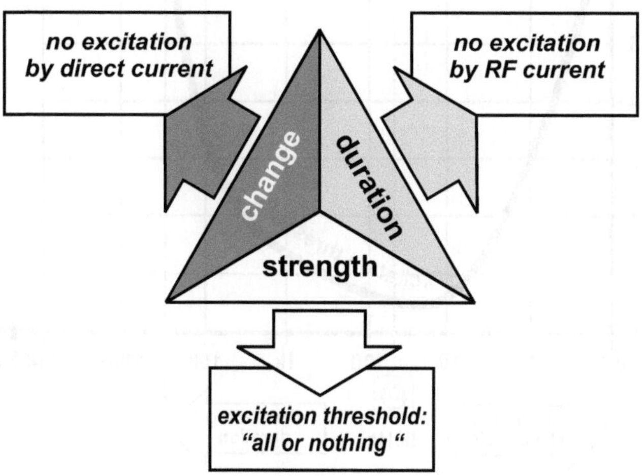

Figure 8-8: The three conditions for cellular excitation: strength, duration and change

8.1.3 Effects of electric currents

8.1.3.1 Alternating current

With sinusoidal currents, the positive (excitatory) half wave is followed by the opposite, the inhibiting negative half wave. Therefore, the stimulus duration is directly dependent on frequency, and it becomes too short for excitation above about 100 kHz. The three excitation conditions lead to the frequency-dependence of cellular excitation characterized by its "bath tube-curve" (Figure 8-9). Unfortunately, both in regard to change rate as well as to duration of the stimulating half-wave the mains frequency meets very well the physiologic requirenments: This makes it stimulating very efficiently. Therefore, hazards from mains frequency electric currents are very high.

Perception thresholds of electric currents are very different from individual to individual. On the basis of investigations with contact areas of 3 cm² the span amounts to two orders of magnitude from 10 μA to 2 mA. Women proved to be more sensitive than men, and children (boys and girls) below an age of 12 years are similarly sensitive to women /49/.

With increasing current sensations become more intensive and painful and muscles contract more and more until they finally cramp. Hand – hand current flow at first results in cramping the arm muscle until we are unable to release a grasped live part (let-got threshold). For 50 Hz currents and exposure time longer than 1 s the let-go threshold for the general population is assumed to be 5 mA, for adults 10 mA

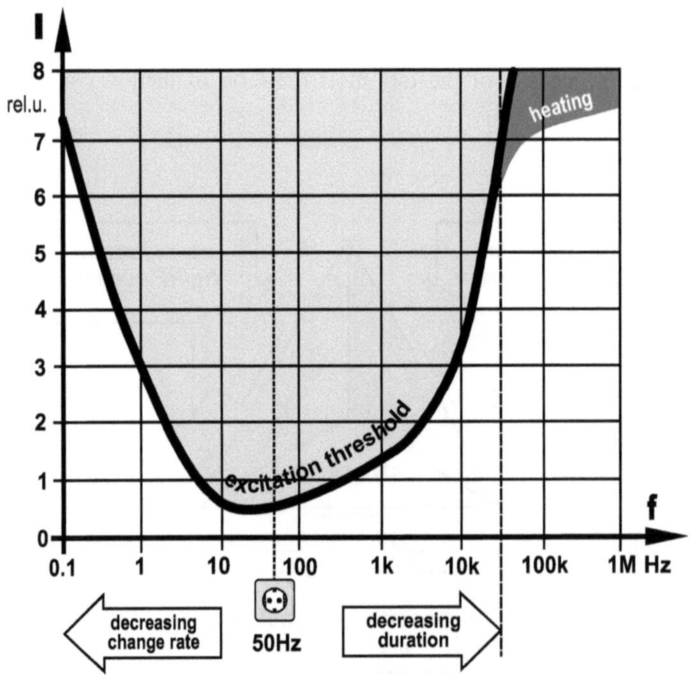

Figure 8-9: Frequency-dependence of the excitation threshold

(EN 60479-1 /26/). With further increasing currents also breast muscles are affected and breathing impaired. Acute danger of life occurs if currents across the heart are strong enough to trigger heart fibrillation. This is expected at about 40 mA. Fibrillation probability increases with further increasing currents to 5% at 50 mA and 50% at 80 mA. This is accompanied by increasing cramping of breathing muscles until apnoea.

Since stimulation depends on stimulus duration, let-go thresholds and perception threshold increase with decreasing exposure time. However, it is remarkable that probability for heart fibrillation decreases at exposure times below 1 s (Figure 8-10), and fibrillation threshold increases up to 4-fold. The reason for this is that fibrillation risk is limited to a short period during a heartbeat, in particular during recovery of the various muscle cells from excitation which for ventricular muscles is at the beginning of the ECG's T-wave (Figure 8-11).

Fibrillation can be explained by the special way in which our heart beat is generated which relies on four special features of the heart:

1. *autonomous excitation:* Our heartbeat is triggered by an autonomous excitation centre, the sinus node which functions independently from external nerval stimulation. This had already been demonstrated by historical reports on Cortéz' conquest of Mexico which described priests sacrificing young men. After removing a victim's

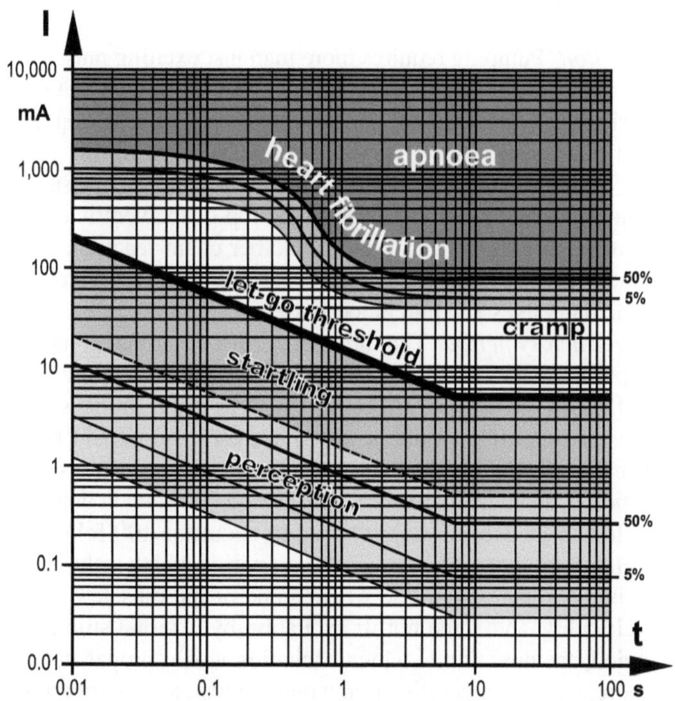

Figure 8-10: Biological effects of electric 50 Hz currents I in dependence of exposure time t for currents flowing from left hand to foot (feet), derived from EN 60479-1 under consideration of perception thresholds of Leitgeb et al. /48/.

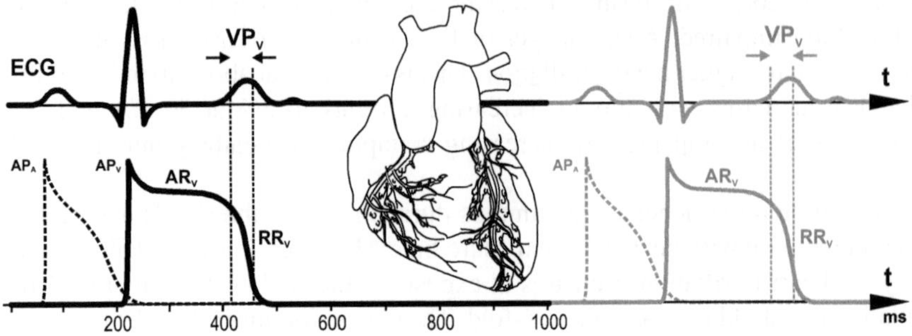

Figure 8-11: ECG and action potentials of atrial cells AP_A (broken line) and ventricular cells AP_V (solid line) with phases of total inexcitability (absolute refractory period) AR_V, phases of re-excitability (relative refractory period) RR_V and vulnerable phases VP_V.

heart from the body, it still continued beating in their hands – as a consequence of its autonomous excitation.

2. *domino effect:* Heart muscles differ from skeletal muscle by an important ability. When excited, they are able to directly pass on their excitation to neighbouring cells. Therefore, the trigger generated by the sinus node is sufficient to generate an excitation wave propagating across the heart's atrium muscles in a chain-reaction like a domino effect. Consequently, atrial convulsion pumps blood from atria into the ventricles.
3. *coordinated action:* Pumping requires more than just exciting muscle cells but also coordination of actions to allow blood to flow into the ventricle. Only after atrial activity ventricular muscles should compress from top to base to pump blood into the aorta and pulmonary artery. This is assured by avoiding the passing over of atrial muscle excitation to ventricular muscles through electric insulation. Instead, it is picked up by nerves and transferred to the top of the ventricles where it is released to ventricular muscles and triggers another chain-reaction. This causes another excitation wave propagating across ventricular muscles which presses blood against ventricular valves and into blood vessels.
4. *prolongated refractory period:* The three properties described so far already allow coordinated pumping – as long as excitation waves propagate in the intended direction only and not in the reverse. However, the weak point in this system would be cellular recovery time. If it would be too short, excitation could be passed backwards, coordination would be lost and pumping would stop. To assure that the intended direction is maintained the refractory period (the time in which heart muscle cells can no longer be excited) is prolonged to about the 100-fold, compared with nerve cells. However, there still remains a short period which is the Achilles' heel of the heart. It is the vulnerable phase where some of the muscle cells have already regained excitability, and some not. If excitation by external stimuli would fall in this very period, the direction in which excitation would propagate cannot be predicted anymore, excitation coordination would be lost, and the heart muscle would start trembling (and stop pumping) without any chance to return to the initial status. Consequently blood circulation ceases – with lethal consequences within a few minutes.

If short pulses stimulate after the vulnerable phase when all muscle cells have already regained excitability, this triggers only an additional heart beat (an extra systole). If stimuli occur before this, in the absolute refractory phase, cells are unable to respond and nothing happens. Therefore, if exposure to electric currents is smaller than the duration of a heart beat, fibrillation probability decreases.

Fibrillation cannot be found in ventricular muscles only. Also atrial muscles exhibit a vulnerable phase with risk of fibrillation. However, while ventricular fibrillation is lethal, atrial fibrillation may only reduce the heart's pumping efficiency.

The only way to terminate fibrillation is to apply a very strong stimulus which is able to stimulate every excitable heart muscle cell. This terminates uncoordinated activity and allows the sinus node to start again triggering excitation waves. However, in case of atrial fibrillation with still pumping ventricles, defibrillation impulses may end atrial fibrillation but initiate ventricular fibrillation if delivered in the vulnerable phase. Therefore, atrial defibrillation requires appropriate timing. This is possible by ECG-triggered release (defibrillator's cardioversion option).

8.1.3.2 Direct current

It is the nature of direct current that it remains constant and, therefore, is unable to stimulate cells. However, in spite of this it is not harmless. This has two reasons:

1. in contrast to alternating currents where charges oscillate around their rest positions, direct currents cause charges that continuously move in the same direction from one pole to the other, with positive charges (e. g. H^+ ions) accumulating at the cathode and negative charges (e. g. OH^- ions) at the anode. This affects membrane potentials in terms of cellular potential differences between the negative interior and positive exterior. This does not excite cells but influences their excitability in different ways. Below the cathode the excess of positive charges increase cellular membrane potential, and consequently reduces their excitability, while below the anode aggregated negative charges reduce the membrane potential and, therefore, the excitation threshold. This increases excitability and sensitivity to internal or external stimuli. Their influence on cellular excitability make direct currents applicable for therapy, such as for pain reduction by reducing excitability of nerve cells.

   Remark: Direct current-induced transport of charges can be used in medical therapy to overcome the skin barrier and move drug ions across the skin into subcutaneous tissue (iontophoresis).

2. because of unidirectional charge movement direct currents transport positively charged hydrogen ions (H^+) and negatively charged OH^- ions (Figure 8-12) towards the opposite electrodes (electrolysis). Below electrodes these ions react with chemical counterparts. At the cathode hydrogen H^+ ions react with Cl^- ions to give hydrochloric acid (HCl), and at the anode OH^- ions react with Na^+ ions to give the base sodium hydroxide (NaOH). Since both reaction products (acid and base) can damage tissue, long-term effects of even small direct currents can lead to an inacceptable risk. This is of particular importance for electronic implants (e. g. pacemakers, drug delivery devices) which are powered by DC batteries and exhibit long-term leakage currents.

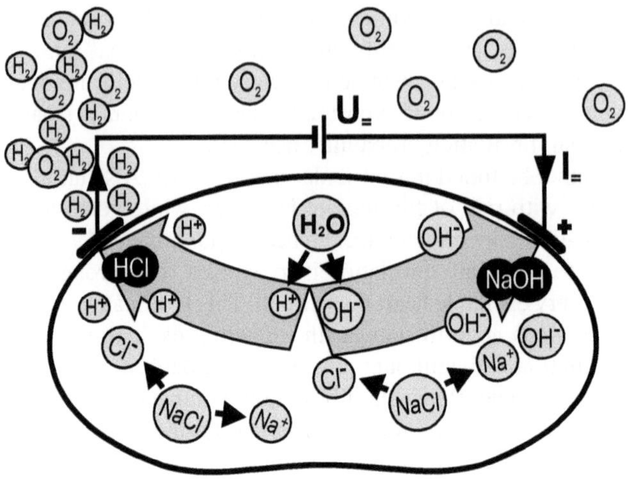

Figure 8-12: Electrolytic effect of direct current on dissociated molecules ($H_2O = H^+ + OH^-$) and generation of acids (HCl) and bases (NaOH) below cathode and anode, respectively. In addition, an explosive mixture of H^+ ions released into air and O_2 (oxyhydrogen) may occur.

3. hydrogen ions that accumulate below the cathode may also be released to the air. This leads to a potentially hazardous mixture of hydrogen with oxygen (oxyhydrogen), which in sufficient concentration can be explosive. This unintended side-effect of DC current flow should be considered (e. g. at hydrogalvanic baths and with iontophoresis but also during RF surgery such as during transurethral resection (TUR) of the prostate where rectification of RF currents due to electric arches may cause DC current components).

Although direct currents cannot excite cells, they may be perceived (e. g. by a feeling of warmth or tingling) above 2 mA (hand – hand pathway). With increasing amplitude, the sensation is intensified and may change to pain. Above 40 mA the propagation of heart excitation might be disturbed. This can become relevant enough to indirectly cause heart fibrillation above 150 mA. Again at exposure periods less than one second the fibrillation threshold increases and stabilizes at 500 mA at durations less than 0.1 s.

8.1.3.3 Heart current factor

If electric currents enter the body they spread over the entire volume depending on the electric tissue conductivity: blood vessels and muscles are preferred, fat, lung and bone avoided. Since the highest hazard is heart fibrillation, it is the amount of current flowing across the heart that has the highest safety relevance (Figure 8-13). It depends on the position of the heart relative to the current pathway. For safety considerations different pathways can be compared in regard to their fibrillation risk by the heart current factor f_H (Table 8-2). It is determined relative to the pathway left hand – foot (feet). Hazard decreases by 20% ($f_H = 0.8$) if currents flow from the right hand to the foot while it increases by 50% ($f_H = 1.5$) for the pathway breast – left hand.

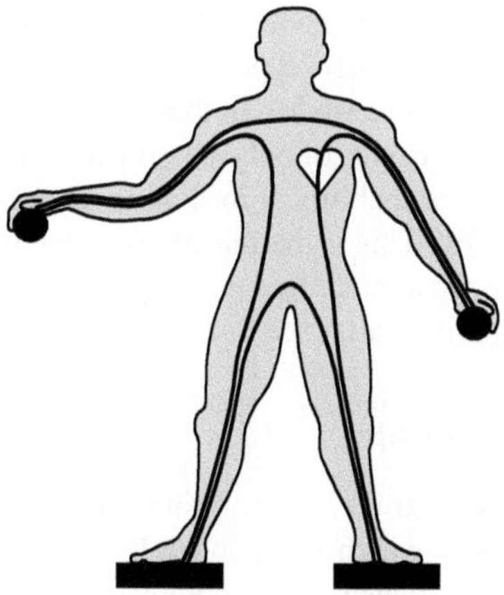

Figure 8-13: Most common pathways in case of an electric accident

Remark: In the case of a thunderstorm the advice is not to lie down on the ground but to try to squat with the feet together. The disadvantage arising from the more unfavourable posture in regard to attracting lightning in comparison with laying flat on the ground is compensated for by two advantages: First, the closed foot position assures that in case of a nearby lightning strike the potential difference picked up by the feet (the step voltage) is much smaller. Second, the heart is additionally protected since only 4% of the step-voltage induced currents are flowing across the heart ($f_H = 0.04$), which considerably reduces fibrillation risk.

In the case of hand – foot pathways heart fibrillation may occur at 50 Hz with currents of about 40 mA. Under more unfavourable conditions (breast – back or breast – hand) already 27 mA may cause fibrillation.

Table 8-2: Heart current factors fH for different current pathways /26/

Current pathways	f_H
Left hand – breast	1.5
Right hand – breast	1.3
left hand – foot (feet)	**1**
both hands – foot (feet)	**1**
Right hand – (feet)	0.8
Left hand – back	0.7
hand (left, right) – backside	0.7
Hand – hand	0.4
Right hand – back	0.3
Foot – foot	0.04

During application of electromedical devices electric currents might be directly introduced into the heart (e.g. applied parts type CF such as cardiac catheters, ultrasonic transducers or RF surgery active electrode). In that case the fibrillation risk is considerably increased. Already small currents of only 10 μA (at contact area 1.2–3.1 mm²) may cause fibrillation with 0.2% probability. Fibrillation probability increases to 5% at 50 μA and 50% at 200 μA /59/. This is the reason, why patient leakage currents of applied parts intended to directly contact hearts (type CF) are set 10-fold below patient leakage currents of other applied parts (see Sect. 8.3.2).

8.1.4 Electric current density

The finding that the electric pathway is important confirms that it is not the electric current but the current density which is responsible for biological effects. Excitation of single cells starts at current densities of about 1 μA/cm². However, conscious perception requires simultaneous excitation of more than one cell and starts at about 10 μA/cm² /48/. Thermal tissue damage may be caused by electric current density above 1 A/cm² starting with whitish discolouration which changes to red above 2 A/cm² and then to brown until 5 A/cm². Above 5 A/cm² longer exposure can already cause carbonization.

Remark: The dependence of biological effects on current density (rather than on electric current) is utilized for RF surgery which applies RF currents to cut and coagulate tissue. With small-area "active" electrodes the treatment current is introduced with high density causing mechanical and thermal effects and leaves the body at the large area "neutral" electrodes with current densities small enough to avoid any unintended adverse side-effects. Consequently, unintended neutral electrode loosening (decrease of the contact area) may lead to unintended hazardous tissue burns.

Remark: For conformity classification both active and neutral RF-surgery electrodes are considered "active" and classified into conformity class IIb.

For safety considerations, it is particularly important to know which current densities might cause heart fibrillation. Investigations on the hearts of dogs and different contact areas showed that impairment of pumping may start above 329 μA/cm², and heart fibrillation above 530 μA/cm². However, even lower thresholds cannot be excluded /58/.

8.2 Limitation of Voltages

Electric voltages may not only be accessible under fault conditions. In daily life and in particular in medical locations potential differences are more frequently encountered that we might be aware of. They can have two different causes:

1. in an electric field (such as originating from ceiling lights) every electric conducting object obtains an electric potential depending on its size and position within the

field (capacitive coupling) (Figure 8-17). This is the reason why even between passive metallic objects potential differences (voltages) can be measured. As an example, within an electric field from a ceiling light an isolated equipment trolley may obtain a potential difference to ground of about 700 mV which is considerably higher than the 10 mV limit within the patient environment (Figure 8-15) in class 1 medical rooms (Sect. 8.2.2).

2. according to Ohm's law potential differences are caused everywhere where electric currents flow across resistances. As an example, a (permitted) earth leakage current of 500 μA at the resistance of 356 mΩ of a 30-m protective earth conductor leading from the device to the distribution box causes a potential difference of 178 mV, which elevates the potential of all PE-connected parts of the device accordingly. At permanently installed devices the allowable earth leakage current is 5 mA which would lead to a potential elevation to even 1.78 V. Since usually more than one device is connected to an electric circuit, their earth leakage currents sum up in the protective earth conductor, and consequently even higher potential differences may be encountered which may require additional measures to equalize the resulting potential differences (Sect. 8.2.2).

The question, which voltages must be considered as dangerous, does not only depend on their absolute value but also on body impedance, kind of contact and electric current pathway. In a current-voltage diagram regions of biological effects can be drawn and relationships to body impedances demonstrated (Figure 8-14). It can be seen that voltages which must be considered dangerous can vary in a wide range depending on the particular situation such as encountered in daily life or during cardiac surgery.

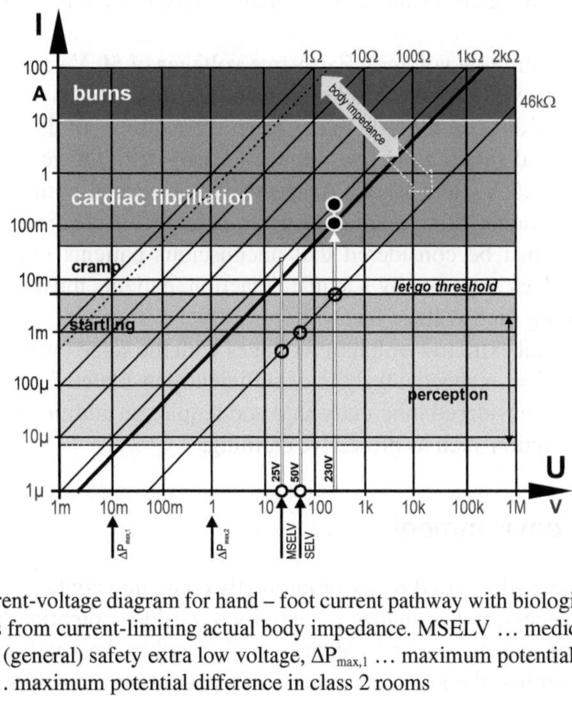

Figure 8-14: Current-voltage diagram for hand – foot current pathway with biological effects. Danger of voltages results from current-limiting actual body impedance. MSELV … medical safety extra low voltage, SELV … (general) safety extra low voltage, $\Delta P_{max,1}$ … maximum potential difference in class 1 rooms, $\Delta P_{max,2}$ … maximum potential difference in class 2 rooms

Since already 27 mA can cause cardiac fibrillation in unfavourable current pathways (e.g. breast – back, or breast – left hand with a thorax resistance of about 10 Ω) already 270 mV could be hazardous to life.

The probability of cardiac fibrillation caused by currents of 10 μA directly introduced to the heart is about 0.2%. Since the electric resistance of the heart is only a few Ohms, directly coupled voltages of only several 10 μV could be dangerous. In case of indirect hand – hand contacts and a body resistance of 1,000 Ω potential differences of about 10 mV could already cause cardiac fibrillation. For these reasons, potential equalization measures must be taken to avoid potential differences larger that 10 mV within the patient's environment in operating theatres intended for critical surgery (e.g. cardiac surgery).

! Already 10 mV can be dangerous to life

Therefore, if in operating theatres for heart and thorax surgery the maximum potential differences are limited to 10 mV this does not reflect dramatically increased safety but just the attempt to keep the generally accepted safety level even under such particularly unfavourable circumstances.

8.2.1 Safety voltages

The body impedance depends on contact area, contact site and the current pathway across the body and may vary between about 10 Ω and 40 kΩ. To guide design and assessment of technical safety means safety voltages have been defined as follows:

- **SELV** (safety extra low voltage): Earth-free voltages of **50 V~** and **120 V=**, respectively, generated with double insulation from (earthed) mains voltage are considered safe even if directly touched under general conditions of daily life.
- **MSELV** (medical safety extra low voltage): Earth-free voltages of half the SELV values, namely **25 V~** and **60 V=**, respectively, generated with double insulation from (earthed) mains voltage are considered safe also in medical locations. However, since it must be considered that unconscious patients may not be able to protect themselves anymore by adequate reaction, patients must be protected from directly touching such voltage levels.
- **FELV** (functional extra low voltage): Voltages with the same level as SELV, namely **50 V~** and **120 V=**, respectively, but not sufficiently insulated from (earthed) mains voltage are not considered inherently safe and require an additional means for electric shock protection such as protective earthing.

8.2.2 Patient environment

If there is a risk for patients to directly or indirectly come into contact with critical voltages, means are necessary to equalize potentially hazardous potential differences. This is considered necessary in medical locations (only) within the "patient environment." This is that zone around the intended position of a patient during diagnosis and therapy where simultaneous contact to objects and patients can occur via the user. In this way patients could unintentionally be exposed to potential differences (Figure 8-15). On the basis of the average span length of arms the patient environment is defined as the entire volume around the intended position of the patient (e. g. around the operating table) up to a distance of 1.5 m (Figure 8-15). If there is no particular position defined, the whole room is considered the patient environment.

Potential differences may have different reasons:

- earth leakage currents of devices cause voltage drops at the resistances of the earth connection along the way from the device to the socket outlet, sub-distribution box, main distribution box via the grounding electrode finally to earth (Figure 8-16). These voltage drops elevate potentials of various earthed objects depending on their earthing conditions.
  - devices connected to the same electric circuit but at different positions;
  - devices connected to different electric circuits.

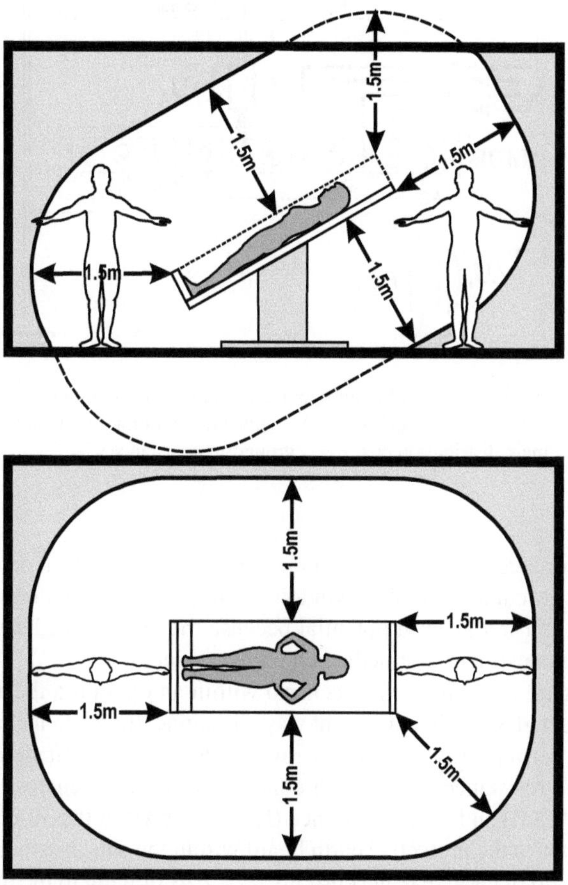

Figure 8-15: Patient environment (where "special potential equalization" is required). Sheer plan (top), and ground plan (bottom) /28/.

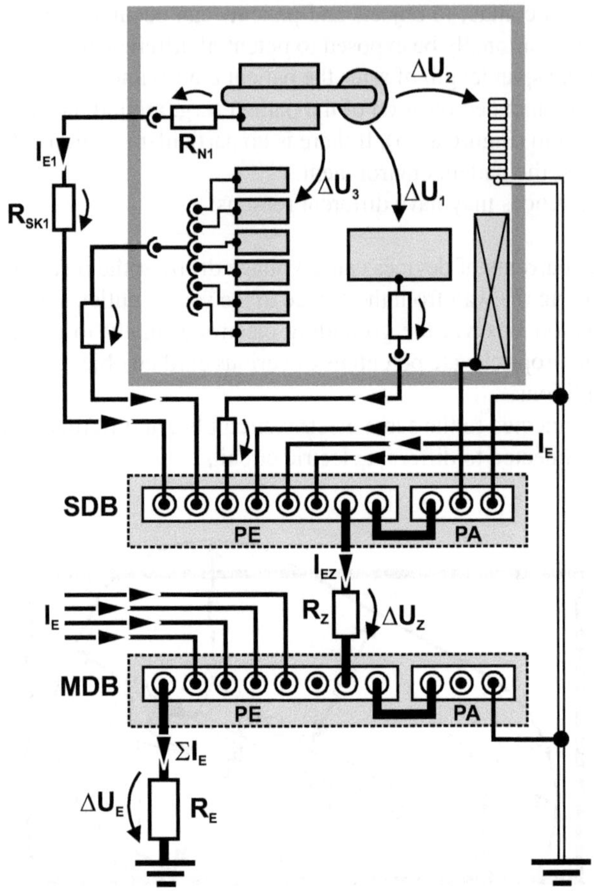

Figure 8-16: Potential differences within an operating theatre due to voltage drops caused by earth leakage currents. PE ... protective earth board, PA ... potential equilibration board, SDB ... sub-distribution box, MDB ... main distribution box, R_E ... grounding resistance

Potential differences (touch voltages) may appear between such devices and metallic objects (e.g. cubicles, tables, window frames, door frames or radiators) which remained at their initial earth potential because they are earthed separately, for example via the construction or dedicated earth conductors.

Example: Figure 8-16 shows the scenario within an operating theatre with various objects connected to earth in different ways. Imagine, the earth leakage current I_{E1} = 5 mA of the permanently installed X-ray device causes a voltage drop of ΔU_{SK1} = 4.3 mV at the resistance of the earth connection from the enclosure to the socket outlet (R_{N1} = 0.3 Ω) and at the resistance (R_{SK1} = 0.56 Ω) of the 30-m long protective earth conductor to the protective earth board within the sub-distribution box. Subsequently, the sum of earth leakage currents I_{EZ} = 10A flowing to the main distribution box causes a voltage drop ΔU_Z = 3 V at the resistance R_Z = 0.3 Ω of the 55-m long earth connector (including lock resistances). Finally, the total sum of earth leakage

currents I_{Eges} = 25 A flowing to ground causes a voltage drop at the earth resistance R_E = 6 Ω of the installation which elevates the potential of the earth board of the main distribution box relative to earth potential to ΔU_E = 150 V.
If no potential equalization at all would be made, within the operating theatre the voltage between X-ray device and radiator would reach the acutely dangerous value of ΔU_2 = 153,0043 V (= 150+3+0.0043 =153.0043 V). Performing the main potential equalization by connecting the heating pipe to the potential equalization board of the main distribution box reduces the touch voltage within the operating theatre to 3.0043 V. But even this voltage would be too high within the patient environment. Therefore, "special potential equalization" is required by connecting metallic objects within the patient environment to the potential equalization board of the sub-distribution box. This allows reducing potential differences to less than 10 mV, namely to 4.3 mV.

- in an electric field such as originating from ceiling lights every electric conducting object obtains an electric potential depending on its size and position within the field (capacitive coupling). Figure 8-17 shows the electric field distribution within an operating theatre by electric field lines and equipotential lines. As an example, the equipment trolley may obtain a potential difference to ground (or the earthed patient) of about 700 mV.

Within the patient environment measurements should demonstrate that potential differences do not exceed permitted levels or/and potential differences need to be equalized by connecting otherwise earthed or floating metallic objects with the potential equalization sockets (which in turn are connected to the potential equalization board of the sub-distribution box) with potential equalization conductors (Figure 8-18). These detachable potential equalization conductors must have a green/yellow insulation and a standardized plug. It must be emphasized that potential equalization must not be con-

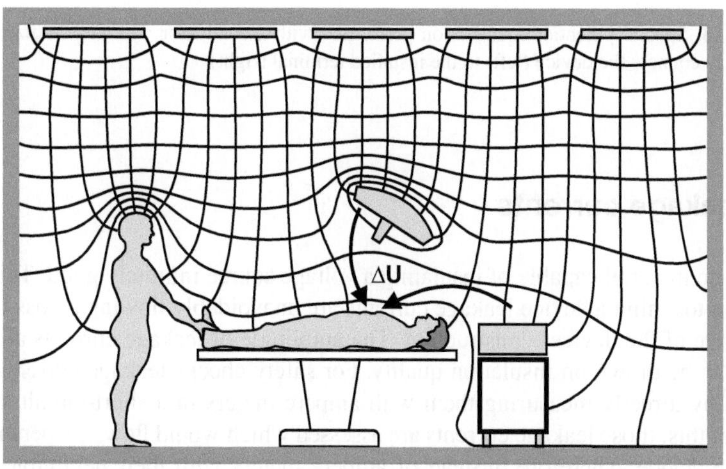

Figure 8-17: Electric field within an operating theatre originating from ceiling lights with potential differences ΔU between insulated electric conducting objects and the earthed patient.

fused with protective earthing. In contrast to protective earth conductors (which are intended to conduct earth leakage and short-circuit currents) the potential equalization conductors are not intended to conduct (relevant) currents.

Remark: Special potential equalization is performed with detachable connections. For this purpose standardized connection bolts must be mounted on class II devices to fit in the standardized plugs of potential equalization conductors. Potential equalization conductors must have a minimum cross-section of 4 mm² (Figure 8-18).

Within the patient environment special potential equalization must involve all floating metallic parts including those of device enclosures and objects grounded by other means such as the operating table, operational lamps, instrument holders, supports, window frames and door frames.

Remark: Already in the planning phase and during construction provisions must be made allowing special potential equalization of fixtures such as window frames, conducting floor layers or built-in cubicles.

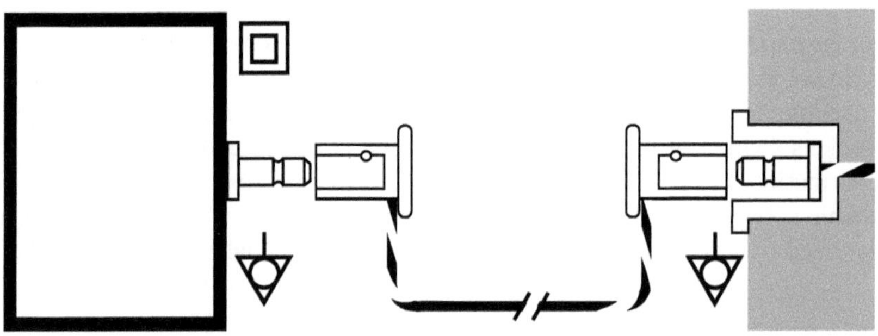

Figure 8-18: Special potential equalization performed with a conductor with standardized plugs at both ends to connect the device (left) to the installed terminal (right)

8.3 Leakage currents

There is no material capable of insulating a voltage source infinitely good. This means that when touching a device leakage currents are unavoidably flowing across our body irrespective of the devices construction. The amplitude of leakage currents allow conclusions to be drawn on insulation quality. For safety checks leakage currents are not assessed by directly measuring them with ampere meters in a short-circuit situation. Instead of this, those leakage currents are assessed which would flow, if a person would touch the device. Therefore, instead of ampere meters with their negligible internal impedance an equivalent circuit is applied mimicking the patient in terms of its body impedance and frequency-dependent excitation (Figure 8-19).

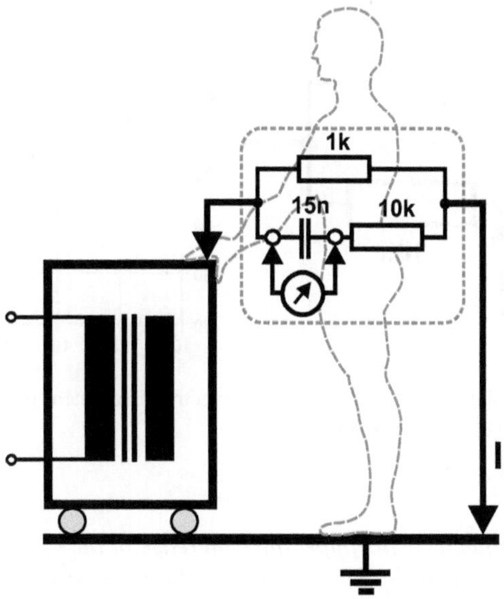

Figure 8-19: Measuring the leakage current from enclosure to earth potential (touch current) with a patient-simulating measurement circuit

For medical devices the electrotechnical protection goal is limiting leakage currents flowing across the insulation to safe values both during normal condition of intended use as well as during single fault condition.

!

Protection goal is limiting leakage currents to safe values under normal and single fault condition

The patient-simulating circuit consists of a parallel circuit of a 1k Ω impedance and an RC-serial circuit consisting of a 10 kΩ resistance and a 15 nF capacitor (Figure 8-20). The leakage current is measured via the voltage drop at the capacitor with the equity 1 μA ≙ 1 mA.

The frequency-dependent measurement of impedance is not intended to mimic the patient's whole-body impedance with the frequency-dependent contribution of the skin impedance since patients are considered as having injured skin. The objective is to simulate the frequency dependence of cellular excitation (Sect. 8.1.2). Since excitation thresholds increase with frequency, this would permit also increasing amplitudes of electric current's frequency components with higher frequency (e. g. harmonics). The measurement circuit (Figure 8-20) allows this because the impedance decreases with increasing frequency which requires higher current amplitudes for measuring the same voltage drop. This allows spectral weighting of non-sinusoidal leakage currents without dedicated spectral analysis. Multifrequency leakage currents may be caused for example by non-linear electronic components or by phase-clipping electronic power control.

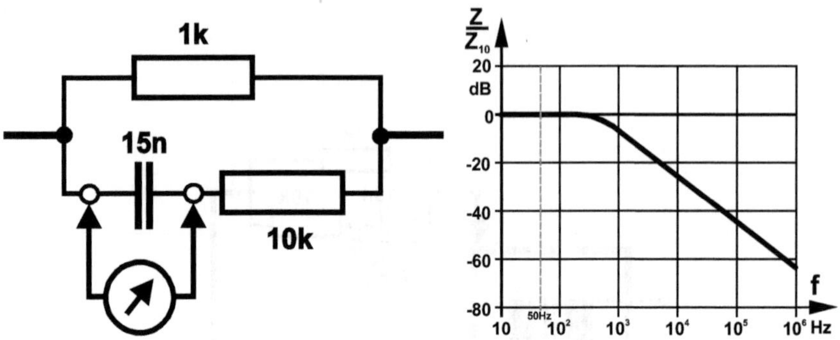

Figure 8-20: Patient simulator for measurement and spectral assessment of leakage currents (left); frequency response (right) (§ 8.7.3 EN 60601-1)

Remark: At low frequencies the impedance increase of the capacitor makes currents across this shunt negligible. Therefore, the voltage measured at the capacitor equals the voltage drop at the 1k Ω resistor which mimics the internal body resistance. With increasing frequency the impedance of the capacitor declines and more current is needed for the same voltage drop. This allows higher amplitudes at the same voltage limit.

Limitation of leakage currents under normal and single fault condition and (failure currents) has the following objectives:

- limiting ***leakage currents*** for normal conditions has two objectives. On the one hand, this should protect from adverse effects when touching devices under normal condition. On the other hand, leakage currents are a diagnostic parameter for assessing insulation quality. Therefore, leakage currents in excess of limits are not only assessed in regard to whether they have reached dangerous levels but, more importantly, they are interpreted in regard to whether the insulation is still sufficient.

  Remark: Elevated leakage currents must not be accepted by the argument that they are still well below dangerous levels. Instead, if the limits for leakage currents are exceeded it must be interpreted that insulation has been severely degraded.

- limits for single fault condition should protect from hazards if a failure has already occurred such as defective insulation. Therefore, leakage currents in single fault condition indicate the quality of failure protection.

Assessing electric insulation by measuring leakage currents has replaced historical approaches of direct impedance assessment with voltage/current measurements. Since the impedance measurement voltage 500 V was potentially harmful to devices this alternative approach should prevent from such unintended side-effects.

Remark: Direct impedance measurement is now restricted to such cases where recurrent testing of in-use devices results in doubts whether the insulation quality might still be sufficient (Chap. 10.10.1.3).

Safety standards do not restrict investigating the insulation of accessible enclosures from mains voltage only. Moreover, electromedical devices also require assessment of insulation of parts that may contact the patient and insulation to earthed parts, independent of their accessibility. Consequently, electromedical devices require measuring the following leakage currents under normal and single fault condition:

- earth leakage current;
- touch current (in the past: enclosure leakage current);
- patient leakage current.

  Measurements are performed during standby and during full operation of devices.

8.3.1 Touch current

Touch currents (previously known as *enclosure leakage current*) are those currents that flow from an accessible part across a person either to earth or to another part of the device (Figure 8-21). Therefore, these leakage currents allow assessment of the insulation between live parts (not just the mains part) and the user.

Remark: In the 3rd edition of the generic medical devices standard /42/ the term "touch current" replaces the term "enclosure leakage current" as used in former editions. In parts two of EN 60601 which are for particular devices the former term is still in use.

For ***household appliances***, the expected contact of users is restricted to the enclosure. Since persons are considered conscious and able to react to perceived electric currents it is sufficient to limit enclosure leakage currents to protect from acute hazards. Therefore, enclosure leakage current limits are higher (EN 60335-1 /25/) and amount to

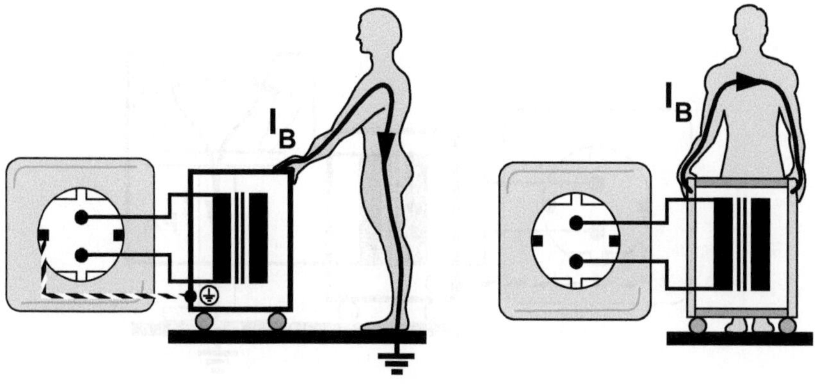

Figure 8-21: Touch current I_B flowing from enclosure to earth or between parts of the device

3.5 mA for permanently installed devices (e. g. electric range or washer). The safety factor to dangerous effects such as to the let-go threshold which is 5 mA for the general population /26/ is only 1.5-fold. Perception of current flow is accepted in those cases where startle reactions (electroshocks) are expected to have no indirect adverse effects. This is assumed with floor-mounted and permanently installed devices but cannot be extended to handheld devices (e. g. drills, mixers). In that case the lower limit of 0.25 mA for enclosure leakage currents should also prevent from the danger of startle reactions which may have subsequent consequences (e. g. fall from a ladder). However, recent investigations showed that such a limit prevents only 50% of the population from perception of the currents /48/, /49/.

Remark: In daily life the protecting impedance is usually considerably higher than assumed for safety considerations because of additional contributions from stockings, socks and shoes. Their additional impedances reduce the actual touch currents amplitudes and increase the safety margin.

Electromedical devices need increased insulation from live parts, in particular from mains voltage, compared to household appliances (Figure 9-2). This is reflected by a lower touch current limit which is 100 μA (AC) /27/. The reason is that within medical locations additional hazards might be associated with too high touch currents:

- startle reactions (electroshocks) could trigger dangerous involuntary movements of the user with severe consequences for example during surgery or treatment.
- indirect coupling of currents into the patient could interfere with electronic implants or even cause heart fibrillation.

8.3.2 Patient leakage current

Patient leakage currents are those that flow from applied parts (which intentionally are brought into contact with the patient) across the patient to earth (Figure 8-22). This leakage current allows assessment of the insulation between live parts and applied parts.

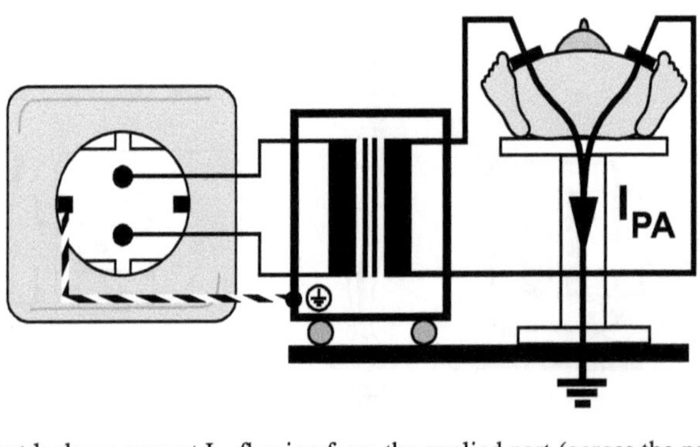

Figure 8-22: Patient leakage current I_{PA} flowing from the applied part (across the patient) to earth

Applied parts (e. g. ultrasonic transducers or biosignal electrodes) can contact the patient's skin only or even be brought into the body's interior via natural or surgical orifices (e. g. endoscopes, catheters or surgical instruments). An even greater protection is required if applied parts are intended for direct contact with the heart (e. g. intra-surgical ultrasonic transducers, intra-cardial catheters).

In normal condition patient leakage currents are limited to 100 μA (AC). For applied parts which are intended for cardiac application they are reduced by 10-fold to 10 μA (AC). Even this is a compromise since in that case the residual fibrillation risk is still 2%. Because of the reduced safety distance to danger any excess of CF-applied part limits must be critically assessed. For single fault condition limits of patient leakage currents are increased by 5-fold.

For devices with several applied parts the overall patient leakage current with all applied parts of the same type connected together must not exceed single part limits in normal condition more than 5-fold and single fault condition more than 2-fold.

8.3.3 Patient auxiliary current

In addition to unintended patient leakage currents there may be another kind of current intentionally flowing across the patient, in particular from one applied part to another. These auxiliary currents are not intended for treatment nor to cause biological or therapeutic effects. They are just necessary to enable the functioning of a method or device such as for assessing body impedance from measured currents and voltages (e. g. impedance cardiography, impedance plethysmography or impedance imaging).

Patient auxiliary currents are measured between applied parts (Figure 8-23). They are limited to the same values as patient leakage currents.

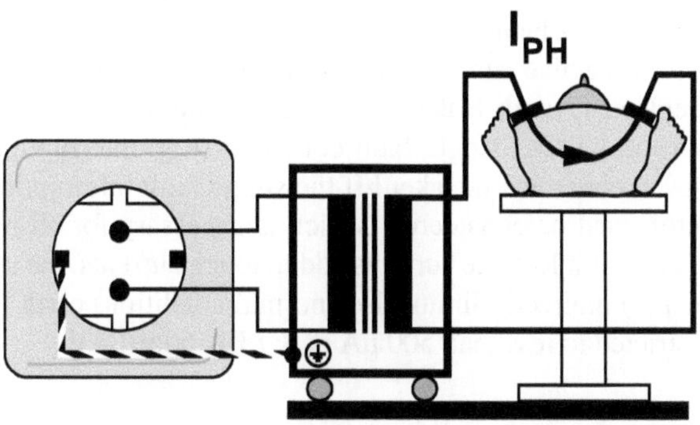

Figure 8-23: Patient auxiliary current I_{PH} flowing across the patient between applied parts

8.3.4 Earth leakage currents

Earth leakage currents are those that flow from live parts, in particular the mains part, across the insulation to earthed parts and subsequently along the protective earth conductor to the installation. Therefore, such leakage currents only exist in safety class I devices.

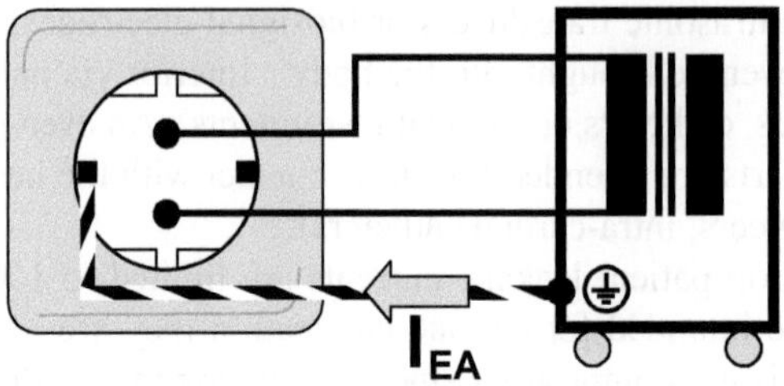

Figure 8-24: Earth leakage current I_{EA} flowing from mains part to earthed parts

The measurement of earth leakage currents allows assessment of insulation from live parts to earthed parts. This is important for two reasons:

- in *normal condition* earth leakage currents flow within the protective earth conductors. Therefore, they cannot directly affect persons. However, along protective earth connections they may cause voltage drops and subsequently elevate electric potentials of protectively earthed parts. Potential differences within the patient environment in the operating theatre could reach values (e. g. above 10 mV or 1 V) which might be harmful to the patient (Sect. 8.2.2).
- in *single fault condition,* when the protective earth conductor is interrupted, earth leakage currents become directly safety relevant. Since backflow along the conductor is made impossible, earth leakage currents add to the touch current and can lead to excess of single fault condition touch current limits.

Interruption of protective earth conductors must not be considered a single fault in any device. This single fault can be neglected in case of permanently installed devices, or of a second redundant protective earth conductor. If interruption of protective earth conductors can be neglected, earth leakage currents are limited to 5 mA under normal condition and to 10 mA under single fault condition (e. g. interruption of the neutral conductor). However, care must be taken! If the single fault interruption of the protective earth conductor must be considered (which is necessary for all non-permanently installed devices) and earth leakage currents add to touch currents, the single fault touch current limit (500 μA) indirectly limits also (normal condition) earth leakage currents which are now restricted to less than 500 μA (§ 8.7 EN 60601-1)!

8.4 Basic assumptions in safety technology

Electric voltages are very beneficial since they feed our electric appliances, but the other side of the coin is that they are dangerous and, therefore, require careful protection. For safety considerations in safety technology the following basic assumptions are made:

- persons (patients and users) are assumed to be ***well earthed***. Insulating clothing, stockings or shoes as well as insulating floor coverings are neglected.

Remark: The assumption of sound earthing is conservative. Already floor coverings exhibit a high insulation resistance (e. g. PVC-floors $R > 10^8\,\Omega$), and shoes have resistances well above our whole body impedance (e. g. leather shoes $R \geq 15\,k\Omega$, shoes with rubber soles $R \geq 10^6\,k\Omega$).

Remark: In exceptional cases outside the field of medical technology the insulation of a person's location is accepted as a safety measure.

- ***users*** are assumed to be protected by their dry and intact skin which increases the body resistance to 2 kΩ.

  Remark: This assumption (together with awareness and reactivity) is the reason why the 3rd edition of the medical device standard lowers the degree of protection for users compared to patients.

  Remark: This assumption is not justified where wet skin needs to be assumed such as in hydrotherapy rooms or bathrooms.

- patients are assumed to have ***less protection*** because their skin might be injured and/or wet. Therefore, for safety considerations it is assumed that protection is based on the internal hand – hand body resistance (1 kΩ) only.

  Remark: A hand – hand contact is the most frequent scenario in daily life, but not necessarily for patients in medical locations. The worst case would be a patient lying with the back on the earthed operating table and coming into contact with voltage at the breast or even within the thorax. In such a case the protecting body resistance would be decreased by more than two orders of magnitude to less than 10 Ω.

- if a person contacts a live part (e. g. touches a device in singe fault condition) it is assumed that the current pathway is from left hand to foot (feet) and the major part of the current is flowing across the heart (Sect. 8.1.3.3).

  Remark: Other pathways can be considered by applying heart current factors (Table 8-2).

8.5 Safety classes

As already shown, electric 50 Hz currents of only 5 mA (which is about 1% of the current of a 100 W bulb) are already dangerous to healthy people. To keep electric shock risks within acceptable limits, constructive protective means are necessary. According to the principle of double protection, this must be achieved by two independent means to provide sufficient protection also in the case when one of the protection means fails (single fault condition). The safety goal is keeping electric currents flowing across the human body within given limits under normal condition and single fault condition.

Patients should be protected from directly touching live parts unless leakage currents stay within their limits. In general, all live parts need to have at least basic insulation.

Basic insulation

The most important (and most common) safety means is enveloping live parts with protective insulation (basic insulation) for basic protection against electric shock. However, it has to be assured that protection is maintained during the whole expected service life. This means that insulation material must not be unduly degraded by ageing or external influences such as chemical or mechanical stress (e. g. by cleaning and disinfection), heat or radiation (e. g. UV-radiation). However, such requirements are not met by all materials. Even PVC, the most frequently used insulating material, is suitable only to a limited extent. It may lose flexibility and become cracked if exposed to temperatures above 75°C. However, such temperatures are not uncommon within devices. Also surface coatings which may deform due to heating are not suitable for protective insulation.

! Not all insulation material is suitable for protection

Even if plastic is not exposed to excessive heat it may become degraded. An ultrasonic company had to learn this the hard way, when they received complaints about mechanical damage such as cracks in the enclosures of their transducers. The reason was that the used plastic was not resistant enough against disinfectants, and became brittle with time.

Natural rubber degrades with time without specific reasons. It becomes cracked and loses mechanical strength and durability, and finally separates from wires. The insulation capability of wood is unreliable and reduces with humidity. Besides this, wood is flammable and may not be used for fire protection reasons. In addition to these arguments, wood is not suitable for enclosures of electromedical devices because of its surface structure and its pores which do not allow reliable disinfection.

Remark: Household appliances might have wooden enclosures, but only in cases where they are not intended to protect from electroshocks, for instance because live parts are double insulated or there is a suitable internal enclosure.

Functional insulation

Insulation which is intended for maintaining performance rather than for protecting from electric shocks must not meet such strict requirements. For this purpose even surface coatings could be acceptable (e. g. wires of transformer windings).

Additional protection

Additional protection under single fault condition can be achieved by different approaches such as:

- protective earthing (safety class I)
- additional (redundant) insulation (safety class II)
- total separation from mains voltage (safety class battery or internal electric current source, respectively)

It must be emphasized that protection by restricting voltages to "safety low voltages" (safety class III) which is acceptable for household appliances or toys (e. g. electric toy trains) is not considered sufficient to protect patients.

8.5.1 Safety class I *(protective earthing)*

Almost no other safety measure is so trappy and misconceived than applying the safety class protective earthing. Its principle is based on insulating accessible parts from live parts by basic insulation. Under single fault condition, if this insulation is damaged, a protective earth conductor is foreseen to connect accessible metallic parts with earth potential. This should assure limiting touchable failure voltages U_F to a safe level (Figure 8-25). The disadvantage of safety class I is that, unfortunately, just a connection to earth is not sufficient for protection.

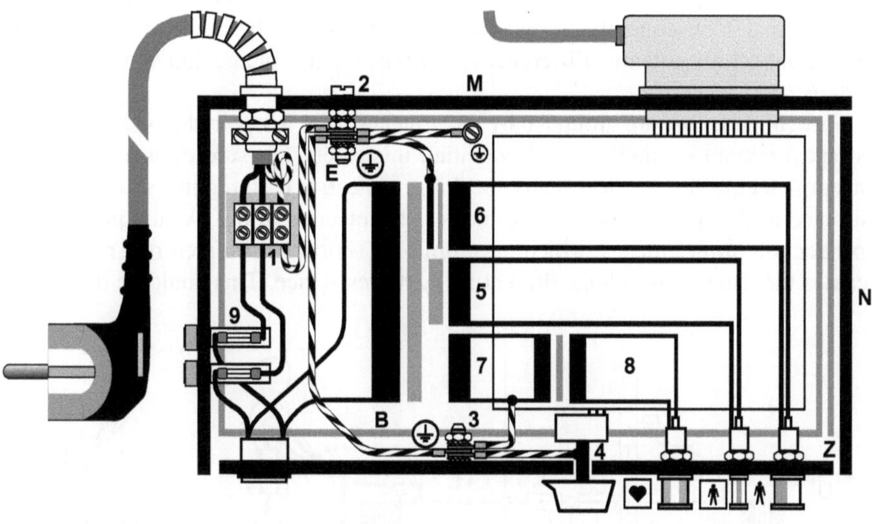

Figure 8-25: Electromedical device of safety class I with basic insulation (B), protective earthing (E) of accessible metallic parts (M) as well as additional insulation (Z) of floating metallic parts (N). 1 … Mains terminal with lagging connection of the protective earth conductor and insulating underlay, 2 … protective earth terminal protected from unintentional loosening from the outside (protective earth star point), 3 … internal protective earth connection, 4 … earthed actuator spindle, 5 … double insulated secondary circuit, 6 … secondary circuit with double insulation and grounded interlayer, 7 … grounded secondary circuit with basic insulation, 8 … output circuit with basic insulation to the intermediate circuit with basic insulation to the mains part, 9 … double fuses

Fire in holiday village

Hall: A short-circuit caused a serious fire within a holiday village. Investigations by fire experts identified a fuse replaced by a nail. This was the reason why the short-circuit current could not be switched off. Excessive heating of the electric circuit ignited the wooden house, and subsequently fire spread over to neighbouring buildings.

This should be explained in more detail. European electric power supply has grounded mains voltage with one of the active leads connected to earth (e. g. the blue insulated "neutral" conductor) and the others exhibiting mains voltage (e. g. 230 V) to earth. Electric circuits are protected from dangerous overcurrents by circuit breakers. The protective earth conductor is connected to the grounding system (e. g. iron bands buried in earth). Therefore, the total earth resistance of a device (to ground) is composed of the resistance of protective earth conductors determined by the conductor length and cross-section, including clamping points, and the contact resistance to ground.

In a single fault condition, when mains voltage contacts the metallic enclosure, at first protective earthing does not protect at all from touching this dangerous voltage. It (solely) causes a short-circuit to ground and a flowing short-circuit current whose amplitude is only restricted by the earth resistance. Therefore, this does not yet limit the touch voltage, but to the contrary makes the situation even worse. The reason is that now the voltage drop at the earth resistance causes a potential elevation of not just the affected device enclosure but also of all other metallic parts connected to the protective earth terminal. Consequently, all these parts (even outside the related room) now exhibit the dangerous potential difference (the whole mains voltage 230 V) to earth (Figure 8-26).

Since all parts that are grounded by other means (e. g. floors, radiators, water pipes, gas pipes) remain on their ground potential, this safety measure even multiplies the danger of electric shock. Electrically speaking, protective earthing just shunts the earth resistance to the body resistance, but does not protect anything. Actual protection is dependent on further means, which must limit the short-circuiting current right before it causes the dangerous voltage drop at the earth resistance. This could be done by an

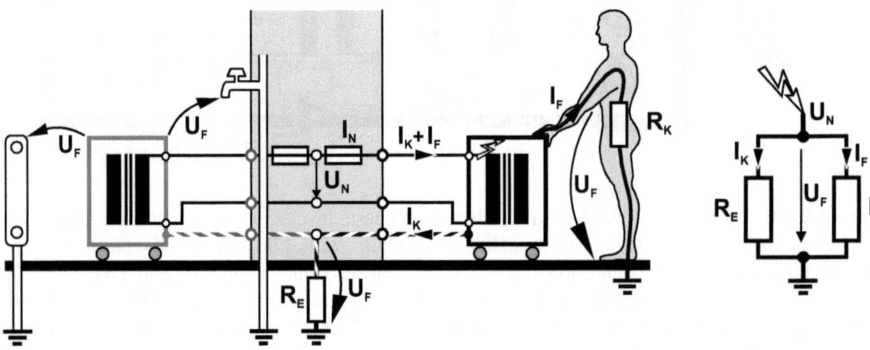

Figure 8-26: Single fault condition and protective earthing (safety class I) with consecutive potential elevation of the protective earth-connected appliances, even outside the related room!

overcurrent circuit breaker with the additional advantage of switching off the circuit and ending the failure condition.

With Ohm's law the value of the touchable "fault voltage" U_F can be determined with the short-circuit current I_k (switch-off current) and the earth resistance R_E to

$$U_F = I_K \,.\, R_E$$

If the protection strategy should be acceptable, the fault voltage must at least not exceed the medical safety extra low voltage $U_{MSELV} = 25$ V~. Therefore, the earth resistance of the electric installation must be low enough to meet the following condition

$$R_E \leq U_F/I_K$$

However, overcurrent circuit breakers usually are not sufficient for protection. The reason is that they of course do not switch off at the rated current but, depending on their characteristics, at 5- to 10-fold the rated current. This means that for a rated circuit current $I_N = 16$ A and a 10-fold switch-off current the earth resistance of the entire electric installation would have to be smaller than 0.16 Ω. Since standards limit already the impedance of the protective earth connector between the device enclosure and the socket outlet to 0.3 Ω, which is about twice the required limit, and the entire earth resistance is about another 10-fold higher, sufficient protection of protectively earthed devices cannot be achieved only by circuit breakers.

The consequence is that protective earthing imperatively requires an additional protective means to limit short-circuit currents to sufficiently low values. The solution is a residual-current circuit breaker that continuously monitors whether currents flowing to appliances duly return in the neutral conductor. If not, this indicates an insulation failure, and the circuit is switched off right before failure currents (and failure voltages) reach dangerous levels. For this reason, safety class I essentially requires that grounded circuits be equipped with residual current circuit breakers that are activated at failure currents of already 30 mA. Such a low failure current would allow earth resistances to increase up to 833 Ω which is well above the usually encountered values.

The installation of a residual-current circuit breaker solves the safety problem of protective earthing, however, there is still a problem remaining. Experience shows that (fortunately) insulation failures are rare. However, infrequent activation reduces movability of the switching mechanism and, consequently, delayed action, which again in single fault condition leads to higher touch voltages and reduced (or lost) safety. To avoid this, it is specified to regularly check residual-current circuit breakers, for example twice a year, to maintain proper operation in single-fault condition.

! Residual current circuit breakers without regular preventive checks may lose their protective effect and critically undermine protective earthing

Within a hospital regular checks are assured by the technical department. However, this is not ensured outside hospitals in medical practices (and in private homes).

Apart from neglect a frequent cause of not checking residual current circuit breakers are the unintended side-effects of interrupting circuits such as the need to cumbersomely readjust all connected electronic watches. However, there are two occasions per year where this needs to be done, anyway. These are the change from summertime to wintertime and reverse. If these occasions are also used to check residual current circuit breakers, two birds can be killed with one stone. We are clearly reminded of the need to perform a check and avoid additional watch adjustments, since we had to do this anyway. Checking is very easy. The only thing needed is opening the distribution box and pressing check buttons of the various residual current circuit breakers (usually more than one). To assure movability, it is recommended repeating this several times to keep the switching mechanism smooth.

Remark: There is no risk for any adverse effect. This operation just connects a resistor from ground to the live conductor which causes an imbalance between currents flowing to and from the distribution box which consequently (should) activate the residual current circuit breaker.

> Don't forget: Changing from summer to winter time (and back) is check-time for residual current circuit breakers

The great advantage of indicating single faults and switching off hazardous situations must be paid for with an important disadvantage. This is the dependency on external conditions. With this safety strategy protection is not only dependent on the proper design of the device (including internal protective earth connections) but also by external connection to protective earth, adequate electric installation, and periodically checking the operation of the residual current circuit breaker.

> The concept of safety class I (protective earthing) is based on periodic testing

The concept of safety class I (protective earthing) is based on periodic testing (see Chap. 10.10) which comprises

- checking internal protective earth connections and the external protective earth connector of the mains cable by measuring protective earth impedances /33/;

  Remark: Until recently recurrent testing comprised also testing protective earth connections with a high test current (e. g. 25 A) to assure connections could stand short-circuit currents. In the meantime, for convenience of technicians (to avoid carrying equipment with heavy transformers) just measuring resistances with a measurement current (e. g. 200 mA) is considered sufficient (EN 62353 /35/).

- periodic checks of residual current circuit breakers (including repeatedly operating test buttons);
- periodic measurement of release time and release current of residual current circuit breakers.

8.5.2 Safety class II *(protective insulation)*

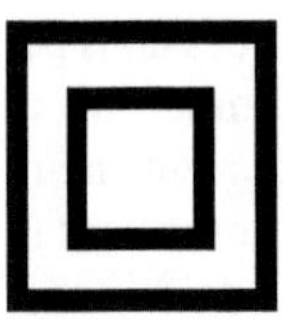

The safety strategy of protective insulation is based on adding another independent second insulation to the basic insulation. Under single fault condition (failure of the basic insulation) this should still provide equivalent protection (§ 6.2 EN 60601-1). The requirement to provide two separate insulating layers instead of only one layer of reinforced insulation should ensure that mechanical damage (cracks) cannot propagate across the whole insulation but stops at the mechanically decoupled second layer. To provide only one single but reinforced insulation layer is only permitted in exceptional cases where two independent layers would not be possible or feasible, such as at the connecting pins of flat mains plugs. An air gap can be a part of the double insulation if it is assured that the separation distance remains unaffected for the whole expected service lifetime (Figure 8-27).

The advantage of double insulation is to provide intrinsic safety, which means that safety of a device does not depend on any additional external condition (e. g. adequate electric installation). Consequently, safety class II devices must not be equipped with a *protective* earth connector. This allows use of such devices also with a less reliable electric installation (e. g. at home or in field hospitals). Intrinsic safety is the reason why double insulation is preferred for handheld appliances and kitchenware (with the exception of irons). Protective insulation is also common for mains cables.

Safety class II devices can be identified by the standardized symbol and the flat mains cable with the flat mains plug (Figure 8-27).

Remark: Safety class II devices may be equipped with a three-pin plug, provided the earth connection is for functional reasons only (e. g. improving electromagnetic compatibility) and not used for protection. In fact, functional earth conductors of safety class II devices must be treated like live parts and doubled insulated from accessible parts (§ 8.6.9 EN 60601-1).

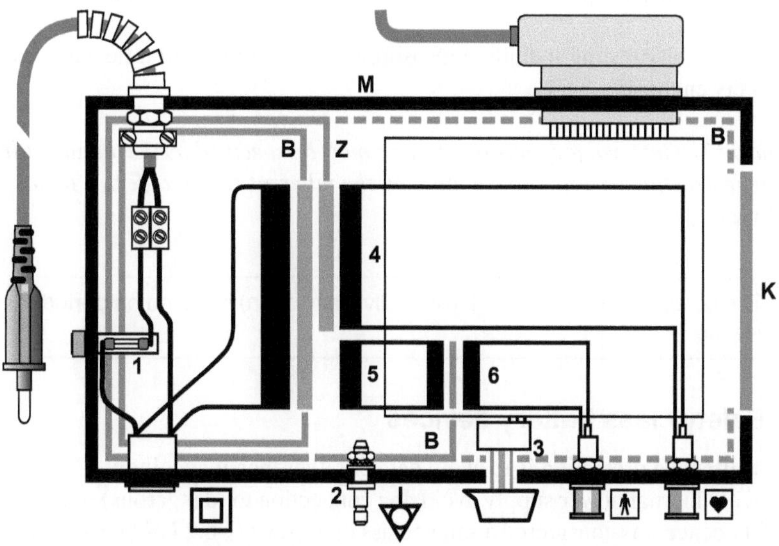

Figure 8-27: Safety class II device with basic insulation (B) of insulating parts (K) and additional insulation (Z) to accessible metal parts (M). 1 ... Single fuse, 2 ... potential equalization connector, 3 ... doubled insulated actuator spindle, 4 ... doubled insulated secondary circuit, 5 intermediate circuit with basic insulation to mains part, 6 ... output circuit with basic insulation to the intermediate circuit (doubled insulation to mains part)

However, the advantage of intrinsic safety and independence of the electric installation must be paid for with an essential disadvantage. Now the single fault condition is neither indicated nor terminated and, therefore, persisting. Consequently, if a damaged insulation remains undetected, the device continues being operated without double protection with the risk that after some time another independent single fault might occur and lead to severe harm. Therefore, insulation damage needs to be taken seriously and professionally repaired.

Remark: Band-aids are not adequate for repairing electric insulation damage, although this is frequently tried in hospitals. On the contrary, band aids are designed so as to have similar pH-values as injured tissue and permit air and humidity to penetrate. Therefore, they are no use as a barrier to electric currents and – which is even more dangerous – mimic protection that is no longer existent.

! Band-aids are no use as electric insulation

The probability of insulation damage is not negligible. It is assumed to be about 1%. For this reason, to maintain safety for user and patients, safety class II requires accompanying measures. They comprise the following activities:

1. visually checking the integrity of insulation by the ***user*** in short intervals. Particular attention should be given to mains plugs and mains cables especially of mobile devices (including hospital beds) where danger of mechanical damage is extremely great.
2. recurrent safety testing including measurement of the insulation (leakage currents) by safety engineers.

Remark: The interval for recurrent testing must be specified by the manufacturer in the instructions for use. Depending on the inherent risk of a device it may be 1–3 years.

! The concept of safety class II (protective insulation) is based on periodic testing

8.5.3 Safety class battery devices

The most effective protection is to avoid danger rather than managing it. Consequently, devices with internal power supply avoid any connection to (dangerous) mains voltage. This safety concept is considered a safety class on its own (§ 6.2 EN 60601-1) provided that devices with internal power supply (battery devices) do not produce dangerous voltages themselves.

However, it is not that simple to define a battery device. Many devices may contain batteries to buffer a blackout, some use mains connections just for recharging batteries and some of them (e. g. defibrillators) provide intended performance optionally either

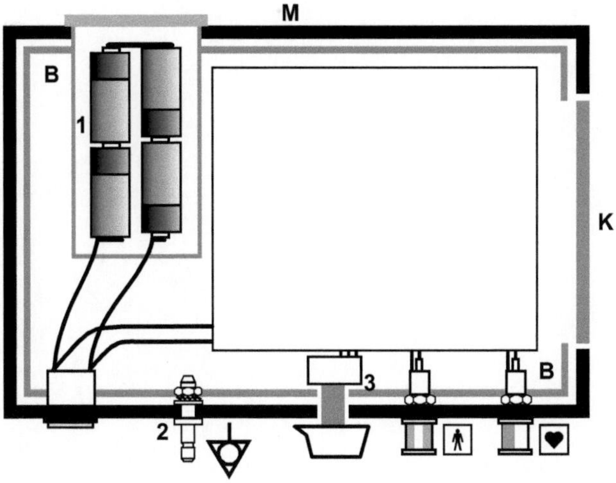

Figure 8-28: Safety class battery device. B … basic insulation, K … insulating part, M … metallic part, 1 … battery box, 2 … potential equalization plug, 3 … basically insulated actuator spindle

with internal batteries or connected to mains voltage. The rule is that devices which during intended use allow connection to mains voltage have to be safety class I or II.

Even for battery devices caution is required not to unintentionally contact patients with dangerous voltages (Figure 8-28). This could be the case if its metallic enclosure would be connected to the applied part. In that case contact with electrostatically charged persons or accidental contact with parts exhibiting excessive potential differences could be transferred directly to the patient. Therefore, applied part circuits must be insulated from the accessible metallic enclosure of battery devices.

9 Electromedical devices

9.1 History of standards

As a general rule, it is considered sufficient if electromedical devices meet those standards which were in force when they were put on the market. It is not mandatory to update devices already in use to meet actual standards. Therefore, for safety assessment of older devices it is helpful to know the historical development of medical safety standards.

Electromedical applications such as electric nerve and muscle stimulation began at the beginning of the 19th century soon after detection of biological effects of electricity by Galvani. The detection of X-rays by Konrad Röntgen in 1896 and the progress in electrical engineering increased diversity and number of electromedical applications.

Remark: Konrad Röntgen had chosen the mathematical symbol for the unknown "X" to name his newly detected radiation "X-rays". Later, to honour him in German speaking countries x-rays were called "Röntgen rays".

In the beginning of the 20th century the first electrotechnical standards were issued, although concentrating on aspects of electrical power supply. At the end of World War II the first safety standards for general electrical appliances were developed, although without special emphasis on medical applications or patient's safety (Figure 9-1). It took until 1977 when the first international generic safety standard for electromedical devices was issued (IEC 601-1:1977 /42/). For the first time, stricter safety requirements were defined to account for the patient's special situation.

Remark: On national level in Germany the first safety standard for electromedical devices VDE 0750-1:1975 was issued in 1975; Austria followed in 1979 with the national standard ÖVE 0750-1:1979.

In 1979 the standard IEC 601-1 was adopted as a European standard (CENELEC HD 395-1:1979 /5/) which harmonized requirements throughout Europe. In several parts two of the generic standard generic requirements were modified and/or expanded for particular device families. All these standards contained specified technical requirements which had to be met by device manufacturers.

In the second edition of the generic standard IEC 601-1:1988 /41/ and its European version CENELEC HD 395-1:1988 /4/ apart from meeting technical requirements manufacturers had already been obliged to perform a risk analysis according to the standard EN 1441 (which in the meanwhile has been replaced by EN 14971) to identify and manage all reasonably foreseeable risks. In addition, manufacturers were allowed to deviate from specified solutions if they were able to realize and prove their alternative had an equivalent safety level.

In the present 3rd edition of the generic standard (IEC 60601-1:2005 /42/ and EN 60601-1:2006 /27/) safety goals were considerably extended, and now include also safe performance, and requirements have been extended to usability and environmentally conscious design and have to consider also foreseeable misuse and human errors (Figure 9-1). In addition, the standard demonstrates a considerable paradigm change. Now device manufacturers are requested not only to perform a risk analysis but to implement and maintain an entire risk management process according to EN 14971:2007. In this process which must be maintained during the whole device life cycle, risk analysis is only one of the elements among others such as verification, validation and continuous post-market surveillance. Now based on own risk assessment manufacturers are entitled to define required safety precautions themselves. This opens the way from strict compliance with specified safety requirements to flexible device-specific application of safety standards, although with the disadvantage that regional and/or individual differences in risk perception and risk assessment might lead to conclusions that are not acceptable by others. In regard to global harmonization this new situation is not necessarily a step forward.

The new leeway given to the manufacturers is a challenge also for notified bodies and safety testers. Now, it is no longer sufficient to check conformity by visual inspection and measurement only. It has become necessary to additionally evaluate whether the direct, indirect and indicative safety means chosen by the manufacturer are sufficient and/or abandonment of additional measures justified and whether the objective of reducing risk to an acceptable level (rather than minimizing risks) is still met. Therefore, also these groups should be familiar with risk assessment und risk management rules.

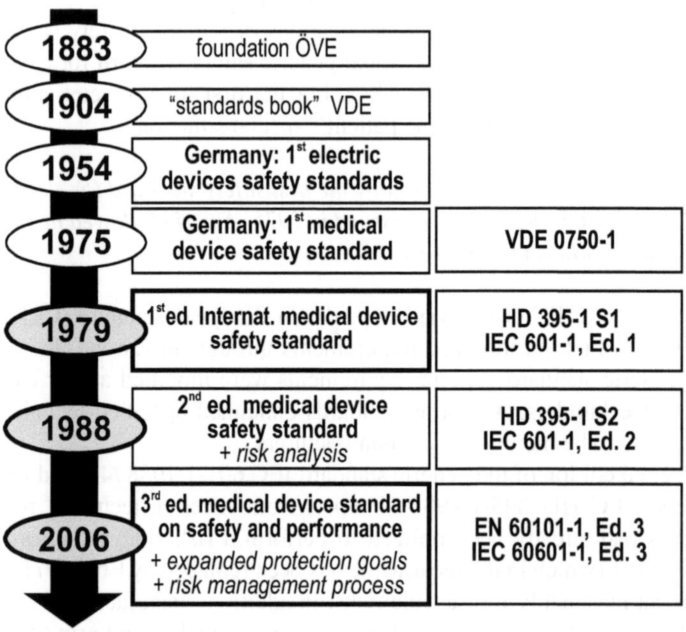

Figure 9-1: History of electromedical device safety standards

9.2 General safety requirements

Safe application of electromedical devices comprises various aspects in regard to the patient, the user and the environment (Figure 9-2):

1. the ***patient*** is in a special situation because of the particular relationship to medical devices through
   - being in direct contact to the electric circuit or even being a part of it and contacted by external skin electrodes or by applied parts introduced inside the body;
   - missing protection from the skin;
   - missing reflexes to reduce or avoid adverse effects due to the patient's condition (e.g. unconscious, unable to move, immobilized) or medication (e.g. analgetics);
   - survival depending on a medical device's function and reliability;
   - impact of imperceptible emissions such as X-rays, (UV-)laser radiation, radioactivity etc.;

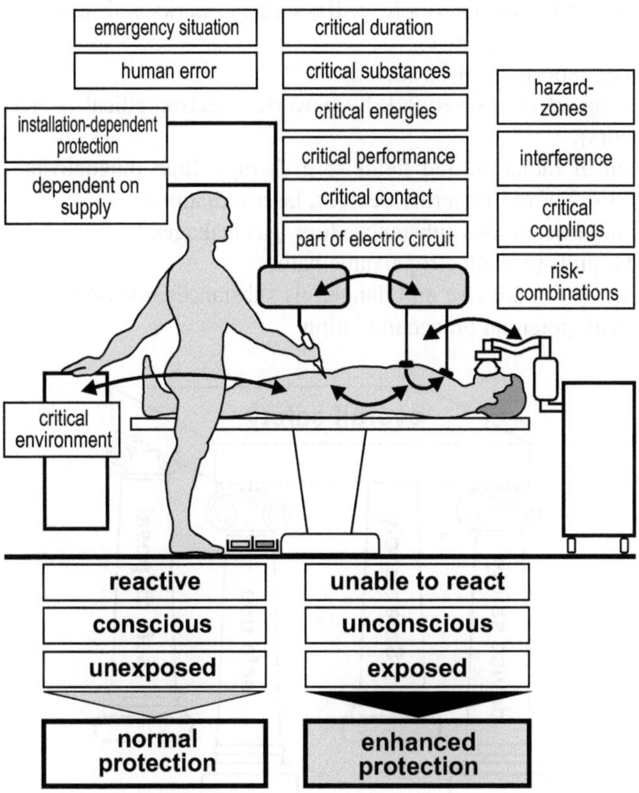

Figure 9-2: Specific risks of electromedical device applications

   - increased infection hazards due to introduction of foreign bodies into the body (e. g. endoscopes, catheters);
   - increased risk of interference and hazardous coupling among devices due to the simultaneous use of several electromedical devices;
   - increased health risks due to released hazardous substances from applied parts;
   - long-term contact with applied parts (or the whole implanted device) and accumulating adverse effects such as tissue damage by electrolytic DC current effects or ageing of materials;
   - exposure to high energies, temperatures, pressure;
   - increased risk of user errors because of emergency situations, stress or tiredness;
   - simultaneous presence of multiple risk factors such as electric energy, high power density and explosive atmospheres or dangerous gases such as endogenous gases or oxygen.
2. ***users*** are exposed to specific risks such as
   - physical factors (e.g. high energies, dangerous invisible radiation, temperatures);
   - dangerous substances (e.g. exhalated anaesthetic gases, leakages of hazardous laser media);
   - pathogens (e.g. direct contamination, inhalation or vaporated tissue);
   - combination of risk factors (e.g. RF surgery sparks and mixtures of explosive gases with air or oxygen).
3. ***environment*** exposed to risks such as
   - disturbance of mains network by powerful electromedical devices (e.g. X-ray generators);
   - emission of radiation and fields (e.g. X-rays, high magnetostatic fields from MRI, RF-EMF of diathermy devices, laser radiation);
   - emission of dangerous substances (e. g. gas leakages);
   - spilled liquids (e. g. electrogalvanic bath).
4. ***ecology*** exposed to adverse and dangerous substances released into ground, water and/or air with potential bioaccumulation.

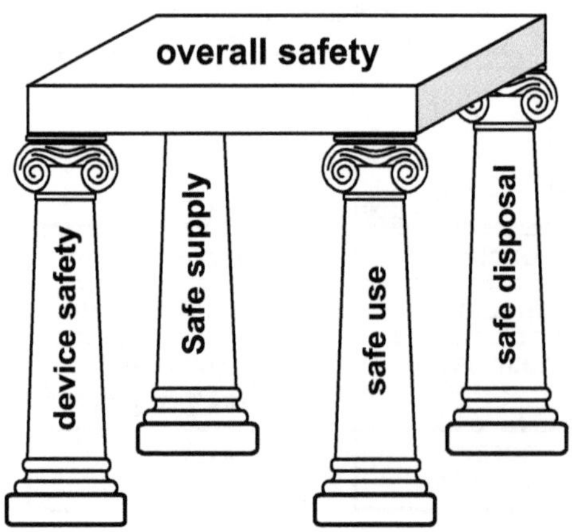

Figure 9-3: Aspects of overall safety of electromedical devices

The overall safety of electromedical devices comprises the following aspects (Figure 9-3):

- ***device*** safety (constructive safety and functional safety);
- safe ***application*** (with consideration of user's knowledge, foreseeable errors and misuse);
- safe ***supply*** (safe electric installation and reliable power supply);
- safe ***disposal*** (environmentally conscious design and waste management).

Regulations

Safety requirements for electromedical devices are laid down in the generic standard

EN 60601-1:2006:	Medical electrical devices: Part 1: General requirements for basic safety and essential performance. Supplementary generic requirements to Part 1 (which according to § 1.3 have also to be applied) are contained in the collateral standards
EN 60601-1-2:	Electromagnetic compatibility
EN 60601-1-3:	Radiation protection in diagnostic X-ray equipment
EN 60601-1-6:	Usability
EN 60601-1-8:	Alarm systems
prEN 60601-1-11:	Home use

Remark: The former collateral standards EN 60601-1-1 (medical systems) and EN 60601-1-4 (programmable electric systems) are already incorporated within the new edition No. 3 of the generic standard EN 60601-1:2006.

Additional standards exist for particular devices amending, modifying or replacing requirements of the generic standard. There are already more than 50 part 2 standards available such as EN 60601-2-4: Particular requirements for the safety of cardiac defibrillators.

Standards apply to electromedical devices and medical systems (§ 1) which are characterized by the fact that they are

- directly related to patients (humans or animals), because they have an applied part, transfer energy to or from the patient or detect such energy transfer.
- intended by the manufacturer for medical use in particular
  - for diagnosis, treatment or monitoring
  - for compensation or alleviation of disease, injury or disability.

The definition of the standard differs from the definition of a medical device as contained in the European directive (Chap. 1.2). On the one hand, medical devices are excluded from the scope of the generic standard if they are not directly related to the patient such are sterilizers (used for disease prevention) or in-vitro diagnostic devices such as devices for in-vitro blood gas analysis (used for diagnosis). On the other hand, devices outside the scope of the medical devices directive are included such as devices for veterinary use.

Safety goal

The dedicated safety objective is that medical electrical devices must be free of unacceptable risks (§ 3.10, § 4.2, § 4.7) from

- direct physical hazards (basis safety) (§ 3.10) or
- device's essential performance (§ 3.27)

which may be imposed on patients, users, other persons, the environment and/or ecology

- during intended normal use;
- during manufacturer-defined expected service life (§ 3.28);
- under normal condition;
- under single-fault condition (§ 4.7);
- during reasonably foreseeable misuse (§ 4.1).

Manufacturers are allowed to deviate from specific requirements provided they apply alternative solutions providing the same level of protection and leading to similar or even lower residual risks (§ 4.5).

To meet these general objectives is not as easy as it might appear. As with the small print of a contract which would require increased attention, each phrase must be considered very carefully.

- first of all, it must be clarified what makes a device a ***medical device***. This is not decided by the user and his actual application but by the intention of the manufacturer laid down in the instructions for use. Therefore, a dedicated cosmetic UV lamp cannot mutate into a medical device just because it is used for dermatological treatment.
- in the 2nd edition of EN 60601-1 electromedical devices had to be applied under medical supervision only. The 3rd edition no longer contains this restriction. This has an important consequence: Now the generic standard is applicable also to home-use devices or to devices intended for lay-use (e. g. publicly available automatic defibrillators). This extension of the scope requires appropriate consideration in risk management. In cases where neither adequate knowledge nor training and careful and professional application can be assumed this must be taken into account by fool-proof design and instructions for use. In addition, safety concepts must not rely on safe installation and reliable power supply as required in medical locations.
- manufacturers are not requested to design their devices for any possible use but only for the ***intended use*** as defined in the instructions for use. Therefore, manufacturers may abstain from making a foot switch water-proof if it is not intended to be used in operating theatres, or may abstain from meeting biocompatibility requirements if applied parts are not intended to be brought into direct contact with bare skin or

intracorporal regions. Since the instructions for use contain important information the user is involved in the responsibility for safety.

- depending on the point of view it can be appreciated or regretted that safety must be provided also (or only) under ***single fault condition*** (subsequent failures inevitably caused by a single fault are still considered a single fault). However, the definition of a single fault is very general. It does not only comprise failure of a single (not necessarily electric) protection means but every occurrence of a reasonably foreseeable abnormal condition (§ 3.116). Frequently occurring abnormal conditions are not considered a single fault but already a normal condition (Chap. 2.3.2). It follows from the risk management process whether at all and by which means risks from single faults are controlled.
- in Chap. 2 it has already been emphasized that acceptance of costs for safety is based on a social compromise based on the cost/benefit ratio. The result is that only those risks must be avoided which are ***reasonably foreseeable***. However, it is left open what in detail should be understood by "reasonably" foreseeable. It is obvious that the final decision on whether a particular harm would have been foreseeable is left to the judge. Basically, according to Murphy's Law all potential failures must be assessed in regard to their risk potential. It is allowed that a failure damages a device provided this does not cause direct or indirect danger to others (fail-safe). As an example, device breakdown would endanger patients in the case of life-supporting, life-sustaining or vital monitoring devices. It is foreseeable, that foldable grids of hospital beds might squeeze parts of a body if safety distances are too small or that somebody might unintentionally press against the protruding piston of an infusion pump and deliver a dangerous bolus. It is also forseeable that the connections of gas tubes or batteries could be interchanged, if this is possible through the design. However, it is not required to design every device to be explosion-proof just because it could be imagined that alcohol could be spilt over it, or to increase the robustness of every device so that it withstands a fall from a table just because it could happen; however, this must be assumed for handheld devices.
- risk analysis must consider the entire ***expected service life*** and, consequently, also age- and use-related degradation of components and insulation.
- not only the life and health of patients (humans and animals) must be protected. Safety objectives are also extended to environment and ecology. This also has consequences. It means that unintended release of dangerous substances, radiation or energy must be minimized. As an example, anaesthetic devices must not exhibit relevant leakage of anaesthetic gases, X-ray devices must be equipped with radiation protection filters and shielding, laser devices must have shutters and barriers preventing from unintended emission of laser radiation. Environmental protection requires means of fire prevention such as avoiding flammable material or excessive temperatures to prevent material from melting and flowing or dropping out of the enclosure.

It is one of the basic safety strategies to restrict the effects of single faults to the affected device only and to avoid involving other devices or the whole electric circuit where life-supporting or vital important devices might be in operation. This is the reason why electromedical devices (in contrast to household appliances) must be equipped with internal overload protection means capable of switching off the affected device before overcurrent circuit breakers of the installation are activated and interrupt supply of the entire electric circuit. For the same reason using fuses with rated current values higher than necessary is not allowed as it would undermine this concept.

9.2.1 Device classification

Electromedical devices can be classified according to different criteria, and have to be marked accordingly, in particular in regard to (Table 9-1):

- inherent risks (conformity class),
- method of electric shock protection (safety class),
- ingress protection related to particles and/or liquids (IP-code),
- explosion protection,
- overload protection,
- permitted sterilization procedures,
- permitted operation time.

Table 9-1: Device classification in regard to inherent risks, and protection against electric shock, ingress, explosion, and permitted sterilization and operation time

Risk	Protection against				Operation time	Sterilization procedure
	Electricity	ingress of particles	ingress of liquids	Explosion		
Conformity class	*Safety class*	*IPN_1X*	*$IPXN_2$*	*Degree of protection*		
I (I_M, I_S)	SK I	IP0X	IPX0	–	S1	–
IIa	SKII	IP1X	IPX1	AP	S2	gas
IIb	battery	IP2X	IPX2	APG	S3	chemicals
III		IP3X	IPX3			radiation
		IP4X	IPX4			heat
		IP5X	IPX5			
		IP6X	IPX6			
			IPX7			
			IPX8			

In case of missing specifications the ***default classifications*** of electromedical devices are: safety class I, ingress protection IP20, continuous operation (S1), and no explosion protection.

Risk classes

It has already been discussed in Chap. 1.4.3 that medical devices are classified according to their risk potential (inherent methodical risk, contact, invasiveness) in conformity classes which determine the method of conformity assessment. These classes are

conformity class **I**: no or negligible risk
conformity class **IIa**: small risk
conformity class **IIb**: enhanced risk
conformity class **III**: high risk

Safety class

The number of accepted options for electric shock protection in normal and single fault condition is limited. As discussed in Chap. 8.5 the following choices exist:

safety class **I**: protective earthing (basic insulation and earthing of accessible metallic parts together with residual current circuit breaker and sufficiently quick circuit cut off);

safety class **II**: double insulation (basic insulation and additional insulation);

safety class **internal power supply** (battery devices): earth-free circuits without galvanic contact to mains power system.

Ingress protection

Protection against unintended ingress of solid objects and liquids is indicated by a code consisting mainly of the two letters "IP" (**i**ngress **p**rotection) and two successive numbers N_1 and N_2 (§ 6.3 EN 60601-1) to read

$$\mathbf{IPN_1N_2}$$

The first number N_1 indicates protection against ingress of solid objects of different size. If protection against ingress of liquid should be left open, the number N_2 is replaced by an "X". Protection against ingress of solid objects extends from no protection (N_1=0) until full protection against ingress of microscopic particles (dust-proof, N_1=6) (Table 9-2).

Table 9-2: Protection against ingress of solid objects

Code $\mathbf{IPN_1X}$	**Blocked object**
IP0X	–
IP1X	fist (>50 mm$^{\varnothing}$)
IP2X	finger (>12 mm$^{\varnothing}$)
IP3X	screw driver (2.5 mm$^{\varnothing}$)
IP4X	wire (1 mm$^{\varnothing}$)
IP5X	dust protected
IP6X	dust-proof

The second number N_2 indicates protection against liquids. If protection against ingress of solid objects should be left open, the number N_1 is replaced by an "X". Protection against ingress of liquids extends from no protection (N_2=0) until full protection against ingress of liquids (water-proof, N_2=8) (Table 9-3).

Table 9-3: Protection against ingress of liquids

Code	Protection against	
IPXN$_2$		
IPX0	– (spilling)	– (vertical spillage)
IPX1	dropping water	vertical drops
IPX2	splash water	drops until 15° inclination
IPX3	spray water	drops until 60° inclination
IPX4	splash water	drops from all sides
IPX5	water-jet	jets from all sides
IPX6	high pressure water-jet	jets from all sides
IPX7	immersion-proof	temporal submersion
IPX8	water-proof[a]	permanent submersion[a]

[a]down to an indicated depth

In general electromedical devices must be at least protected against ingress of fingers, particular protection against ingress of liquids is not required (default: IP20). However, if the intended use requires handling of liquids such as immersing sponges into water to contact muscle stimulator electrodes to skin, the device must be protected against dangerous humidification by unintentionally spilled water. The amount of liquid to be considered follows from risk analysis (§ 11.6.3 EN 60601-1). It may be 0.2 l for muscle stimulators or even several litres in case of water electrodes of electrogalvanic baths.

Remark: Ingress of some liquid may be tolerated if it does not increase risk such as by wetting live parts, generate contact to dangerous voltages, or causing malfunction by short-circuiting circuits of relevant device functions.

Explosion protection

Medical electrical devices must not generally be explosion-proof. However, if they exhibit such a protection they have to be marked accordingly. There are two different grades of protection (see also Chap. 6.2.2):

protection degree **AP**: protection against ignition of explosive mixtures with air (symbol: triangle within a green circle);

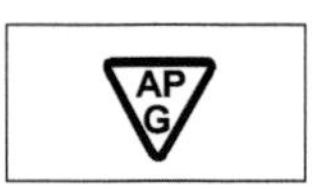

protection degree **APG**: protection against ignition of explosive mixtures with oxygen (symbol: triangle within a green bar).

Operation mode

Operation of devices is unavoidably associated with heating. Usually temperature increases exponentially until it reaches a steady-state value. Insulation of transformers and motors must be designed to withstand thermal load.

Continuous operation (S1): If devices are not specifically marked it is assumed that they are intended for continuous operation.

If the intended operation is time-restricted, the permitted operation cycle must be marked with the graphical symbol at the left showing runtime and interval time indicated in minutes. The following basic options exist:

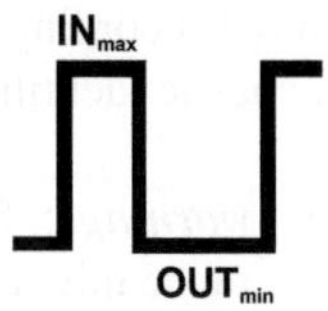

Short-term operation (S2), characterized by interruption before reaching the steady-state temperature and subsequent cooling phases long enough to regain start temperature.

Intermittent operation (S3), characterized by cooling phases not long enough to regain start temperature but sufficient to avoid hazardous temperature increase. In the worst case of load cycles with constant operation and cooling times the temperature exhibits a saw-tooth-like increase with maximum and minimum temperatures approaching steady-state values (Figure 9-4).

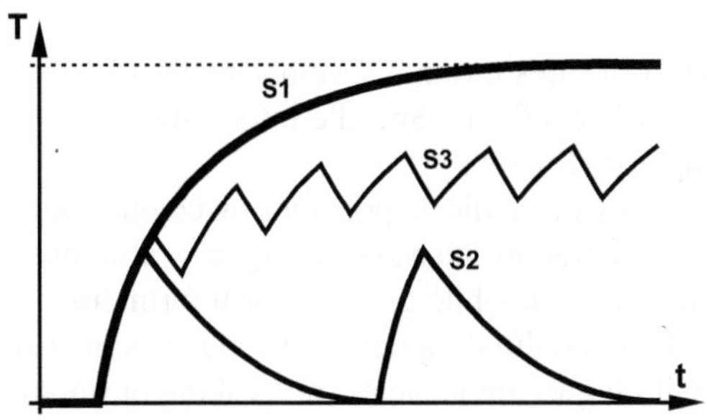

Figure 9-4: Time-dependent heating for continuous operation (S1), short-term operation (S2) and intermittent operation (S3)

Sterilization

Sterilization can affect devices by heating (e.g. destroy piezoelectric ultrasonic transducers), or degrade material properties by chemical reactions (e.g. with disinfectants) or energy quantum effects (radiolysis). For these reasons, manufacturers must specify suitable methods for cleaning, disinfection and sterilization (see Chap. 5). Devices or parts of devices intended for sterilization must be classified accordingly (§ 6.4 EN 60601-1). Packed sterile products must be marked with STERILE. Sterilization methods are added as follows:

STERILE EO sterilized with ethylene oxide

STERILE R sterilized by irradiation

STERILE  sterilized by heating (with a thermometer symbol inside the box)

STERILE .A. sterilized by antiseptic procedures

9.2.2 Alarms

In addition to direct safety by design such as limiting output values to safe levels, the use of alarms for hazardous situations is an important risk management tool. However, because of the numerous patients within an intensive care unit and the increasing number of devices simultaneously in operation even on the same patient it became more and more safety-relevant to differentiate the relevance of alarm signals to inform on their priority in terms of providing assistance. For this reason, based on the supplementary standard EN 60601-1-8 /30/ to the generic standard /27/ alarms have been structured according to their urgency and exhibit a signature allowing immediate optical and acoustic identification. The following alarm classes have been defined:

- ***warnings***: They signal high priority situations requiring immediate action. Ignorance might lead to reversible or irreversible injury or even death of patients.
- ***attention***: They signal situations of medium priority requiring rapid action.
- ***information***: They signal conditions requiring increased attention or precaution but no particular action. Therefore, they have low priority but may be differentiated into critical information and general information.

Labels signal different priorities through specific shapes to allow immediate recognition of priorities (§ 7.5 EN 60601-1). Specific information is given by additional contents or accompanying text (Figure 9-5).

Acoustic and optical signals indicate priorities by colour-coding and time-sequencing. The colour red is reserved exclusively for urgent warnings of acute hazards only and demands immediate action, while green is for information only without the need for reaction (Table 9-4). In addition, optical and acoustic signals indicate priorities by their time signature with higher frequencies for blinking or impulses for higher priorities.

Table 9-4: Parameters of optical alarm signals

Priority	Colour	Blinking frequency	Meaning
high	red	1.4–2.8 Hz	immediate action necessary
medium	yellow	0.4–0.8 Hz	rapid action necessary
low	turquoise	no	no particular action necessary
no	green	no	ready for use
	other colours	no	other information

Figure 9-5: Generic shapes of symbols coding different alarm priorities (from left to right: prohibition, attention, commands, critical information, general information)

Warning labels

Warning labels are related to acute danger. In lack of particular warning symbols (Table 9-5) the general warning symbol should be used and amended by explaining text such as:

FOLLOW INSTRUCTIONS FOR USE!

CONNECT TO ISOLATED GROUND SOCKET OUTLETS ONLY!

DON'T TOUCH PLUGS AND PATIENT SIMULTANEOUSLY!

CONNECTION TO AUXILIARY SOCKET OUTLETS MAY REDUCE SAFETY!

FOLLOW INSTRUCTIONS FOR INSTALLATION!

SITING MUST ALLOW EASY DISCONNECTION!

AVOID VOLTAGE DROPS! CONNECT TO UNINTERRUPT-ED POWER SUPPLY SYSTEMS ONLY!

ATTENTION; IMPROPER EXCHANGE OF LITHIUM BAT-TERIES MAY CAUSE HAZARDS!

IF NOT USED, REMOVE BATTERIES; ACID MAY LEAK!

FOLLOW INSTRUCTIONS FOR STERILIZATION!

FOLLOW INSTRUCTIONS FOR MAINTENANCE!

TILTING DANGER!

DO NOT DISPOSE AS DOMESTIC WASTE!

NO CHANGES WITHOUT MANUFACTURER'S PERMIS-SION!

Table 9-5: Particular warning labels

Symbol	Meaning
	Attention, high tension!
	Attention, fire hazard!
	Attention, explosion hazard!

Symbol	Meaning
	Attention, explosive zone!
	Attention, electrostatic discharges may cause damage!
	Attention, magnetostatic fields!
	Attention, radiofrequency electromagnetic fields!
	Attention, optical radiation!
	Attention, laser radiation!
	Attention, ionizing radiation!
	Attention, substance adverse to health!
	Attention, toxic substance!
	Attention; biological hazard!

Prohibition labels

Prohibition labels demand hazardous activities are not performed. If no particular labels are used (Table 9-6), the general prohibition label should be used and appended with explaining text such as

NOT INTENDED FOR EXPLOSIVE ZONES!

DON'T EXPOSE TO SUNLIGHT!

DON'T OPEN!

DON'T FALL!

Table 9-6: Particular prohibition labels

Symbol	Bedeutung
	Do not reuse!
	Do not resterilize!
	Don't light fire!
	Don't smoke!
	Don't use mobiles!
	No entry for pacemaker patients!
	No entry for patients with metallic implants!
	Don't enter with metallic objects or watches!

Commands

Commands demand particular actions. If no particular labels are used (Table 9-7), the general command label should be used and appended with explaining text such as

CLEAN SKIN BEFORE ATTACHING ELECTRODES!

DETACH MAINS PLUG BEFORE OPENING!

USE PERMITTED ACCESSORIES ONLY!

FIX TRANSPORTATION LOCKS BEFORE MOVING!

Table 9-7: Particular command labels

Symbol	Meaning
	Read instructions for use!
	Wear conducting shoes!
	Wear gloves!
	Wear safety goggles!
	Use respiratory protection!

Instruction labels

Instruction labels refer to correct handling to avoid damage. There is no general generic shape. The objective is to develop self-explanatory labels. Examples are listed in Table 9-8.

Table 9-8: Instruction labels

Symbol	Meaning
	Use by (indicated date)
	In-house use
	Avoid static discharging!
	Attention, fragile!
	This side up!
	Keep dry!
	Protect from direct sunlight!
	Permissible temperature range
	Top-heavy!

Symbol	Meaning
	Don't pile up!
	Don't dispose as domestic waste!

Critical Information

LATEX

Critical information is such that do not require an action but provide important information to trigger awareness for particular risks and stimulate prudent avoidance such as by informing on the content of potential adverse substances such as latex or phthalates.

General information

SN

General information does not require any action but should inform on particular facts. In lack of specific symbols, they are enclosed in a box such as serial number SN, batch number LOT, order number REF, in-vitro diagnostic device IVD, sterility STERILE, control material CONTROL etc.

Table 9-9: Symbols for general information

Symbol	Meaning
	Manufacturer
	Production (indicated date)
	Instructions for use
	Product can be recycled
CONTROL	Control material for performance checks

9.2.3 Applied part

Applied parts are those intended to contact patients; they might have patient connections, or other parts requiring a similar degree of protection, or may occasionally contact patients (Figure 10-3). In regard to electric shock protection applied parts may be earthed (type B) or earth-free (type F) with different degrees of insulation (type BF or type CF), and they may be protected against voltages induced during defibrillation. The degree of protection is indicated by symbols mounted at the connection points on the device or on the applied part, if separately marketed (Table 9-10).

Table 9-10: Marking of applied parts

Symbol	Meaning
	Type B: earthed applied part
	Type BF: earth-free (floating) applied part
	Type CF: earth-free (floating) applied part for cardiac application
	Defibrillator-proof applied part type B
	Defibrillator-proof applied part type BF
	Defibrillator-proof applied part type BF

6.2.3 Applied part

Applied parts are those intended to contact patients; [illegible]

10 Safety testing

10.1 Why testing?

As a basic requirement medical devices must not cause unacceptable risks during the whole intended service life. However, this requirement cannot be fulfilled solely by constructive means but requires involvement of operators and users. The reasons are that several circumstances require recurrent testing such as:

1. the role recurrent testing plays in basic safety concepts is essential. Otherwise, reliable protective earthing of safety class I devices would not be assured, a single fault of double insulation safety class II devices would not be detected, and a battery malfunction of safety class internal power source missed.
2. dangerous material degradation due to ageing and/or abrasion.
3. dangerous degradation of contacts with resulting increase of contact resistances due to corrosion or due to reduced contact pressure caused by mechanical deformation.
4. failure of safety-relevant components (e. g. indicator lamps).
5. safety-relevant device deficiencies caused by erroneous application, error or misuse.

If service life is not explicitly reduced manufacturers are liable for their products and consecutive damages for 10 years after transfer to the client. Since recurrent testing (except for fail-safe design) is an important module of safety concepts, manufacturers are obliged to specify extent and interval of periodic tests in the instructions for use. In turn, operators must follow these instructions. Therefore, before starting testing the instructions for use must be checked for particular testing requirements. Because operators must keep records on medical devices and of the results of recurrent testing, it is common to copy relevant testing instructions and intervals from the instructions for use into the device file at the time of the take-over process.

In spite of the legal requirement to design and produce devices according to the acknowledged state of the art, in spite of due diligence and product liability which in case of consecutive damage could endanger a manufacturer's existence, it still remains the rule rather than the exception that even new devices without third-party testing may exhibit even severe safety problems. The reason is that manufactuer's technicians still concentrate on realizing intended device functions, while knowledge and efforts to meet safety standards are given less weight. Even CE-marking does not assure freedom from deficiencies. Mandatory type testing is only demanded for devices of conformity class IIb and III with increased inherent risks. However, these are the minority among medical devices (Chap. 1.4.5). From this, it follows that devices might not only become degraded by use. Even new devices, in particular of conformity class I and IIa, merit inclusion of detailed visual inspection in the takeover process. This should possibly be done prior to payment to prevent from later problems.

! (Visual) inspection of new devices prevent from later problems

Safety testing of electromedical devices (including receiving tests) is regulated in the standard EN 62353 /33/. It requests safety testing

- as an element of the purchasing process (receiving test);
- periodically in intervals specified by the manufacturer;
- after repair;
- after constructive changes (which were not intended by the manufacturer).

Remark: Changes according to the instructions for use are not considered a constructive change.

Safety testing must comprise the following steps:

1. ***inspection***, comprising external visual inspection. Internal visual inspection has to be added only if requested by the manufacturer or indicated by external clues directing towards potential adverse internal changes.
2. ***measurement*** of safety-relevant parameters such as protective earth resistance (safety class I devices), leakage currents to earth, enclosure and applied parts, and, if degradation is suspected, measurement of the insulation impedance between mains part and the enclosure and to applied parts.
3. ***functional checks*** including measuring safety-relevant output parameters, if applicable.

In addition to recurrent testing by technicians, prior to every application ***users*** are obliged to check the condition of a device. The reason is that it can never be ruled out that something adverse may have occurred even during short periods when the device was not attended, i.e. the device may fall from a table or an object may fall onto the device, ingress of spilled liquid during cleaning or disinfection or overstress of the mains cable or mains plug might have occured. Obvious damage may be identified by the user's visual inspection. If there are doubts as to the condition of a device a technician should be called for safety testing.

Basic principles

Recurrent testing should meet the following basic principles:

1. safety assessment may be performed according to those requirements and standards which were applicable at the time of purchase. It is not required to apply the most recent standard. This means that it is not necessary to continuously adapt devices to the state of the art, unless where the former solution now presents an unacceptably high risk. To avoid unnecessary costs technicians and design engineers should be familiar with the historical development of requirements.
2. test results should be recorded (in the device record). In a fist step, ***all*** identified deficiencies should be listed without regard to their later classification. The reason

is that unlisted deficiencies are considered as not seen. Since risk assessment results might differ among individuals, another person might come to different conclusions.

> ! Not listed means not seen

3. only after their listing must deficiencies be assessed and classified according to their safety relevance.
4. however, assessment rules are not rigid. The safety-relevance of deficiencies depends on the kind of device, its performance, its inherent risks, the circumstances of application and the site of use (e.g. within hospitals or at home).

> ! Risk assessment must consider particular circumstances

10.2 Who is entitled to test?

The market for medical device recurrent testing is huge, and it is possible people without adequate knowledge and training might consider claiming a piece of the economic cake. However, it must be stressed that testing is also associated with responsibility and liability in case of accidents enabled by inadequate testing.

A lethal mistake

Laughing gas instead of oxygen

Innsbruck: A 40-year-old medical assistant died during a spinal disk operation because of interchange of oxygen and nitrous oxide (laughing gas) connections. A technician and an anaesthesiologist were sentenced because of negligent homicide to 9-months conditional imprisonment each.

The technician had poorly performed safety testing and marked the device with "all functions o.k." The anaesthesiologist had insufficiently checked the device prior to application and ignored issued alarms.

Electromedical device safety testing should not only be restricted to just quick measurement of general electric safety parameters but must include also additional testing and checking of safety-relevant performance. Therefore, the required testing effort depends on the kind and intended use of a device.

However, not every technician is authorized to recurrent testing medical devices. The tester must meet the following requirements:

- the required ***legal authorization***.
- specific legal and medical-technical ***knowledge***. Even visual inspection (external and internal) only makes sense if persons are aware of the requirements (laws, ordinances, standards and rules of technology) and know what requires their attention.

The paradigm change in Europe placing emphasis on device-specific risk management also has an impact on medical technology. The importance given to device-specific risk analysis, risk assessment and risk control measures based on a manufacturer's individual judgement now requires also from testing personnel the ability to identify and assess risks instead of just checking conformity with a particular list of requirements. Therefore, additional medical-technical knowledge is required.

- practical ***experience*** in testing, risk identification and risk assessment. The required amount of training depends on the kind and variety of devices to be tested. It is essential that practical experience is gained by testing guided by experienced supervisors.
- required test ***equipment*** must be available, monitored, periodically calibrated and properly documented.
- ***quality management***; requirements on testing and inspection bodies are contained in particular standards (EN 17020 /23/, EN17025 /24/). They demand definition of authorizations and competence, written testing instructions, test equipment recording and quality surveillance including periodic calibration.
- liability ***insurance*** to cover potential claims.

10.3 Device-specific safety goals

The safety goal for electromedical devices is not the complete prevention from access to live parts but to assure that limits for touch currents, electric energies and electric voltages are met under normal condition and single fault condition (§ 8 EN 60601.1). This means that touch protection can also be achieved by high protective impedances. In exceptional cases insufficient air and creepage distances could be accepted if during their short-circuiting safety goals remain met.

Prevalence of non-sinusoidal currents increases due to non-linear electronic components and electronic power regulation by phase clipping. In Chap. 8.1.2 it was shown that biological effects are frequency-dependent. For this reason, leakage currents are not measured just by ampere meters but by a frequency-weighting measurement circuit mimicking a patient's body resistance and frequency-dependent excitability (Chap. 8.3).

All parts, even those hidden behind flaps or covers that are accessible without a tool are considered touchable. A tool is considered any auxiliary means including coins and keys except a part of the body or fingernails (§ 3.127 EN 60601-1).

Remark: All external and internal parts are considered touchable that can be contacted by a standardized test finger. In addition, touch protection is required for parts behind openings of the enclosure that can be contacted by a test pin (15 mm long, 4 mm at its base and 3 mm at its top) and parts which can be contacted by a free-hanging 10-cm test rod through any opening on the top of the device (§ 3.2 EN 60601-1).

To meet the safety goal two separate and equivalent independent safety means must be available (§ 8.5 EN 60601-1). However, now the 3rd edition of EN 60601-1 allows for assessment of these means to differ depending on whether they are intended to protect the user or the patient.

10.3.1 User

The user must be protected against electric shock but to a lower degree than patients. User touch currents are limited to the following values (§ 8.7.4 EN 60601-1):

100 μA in normal condition,
500 μA in single fault condition.

Remark: The earth leakage current is limited to 5 mA in normal condition and 10 mA in single fault condition. However, the earth leakage current adds to the touch current under single fault condition (interruption of the protective earth connector). Therefore, from the requirement to meet the single fault touch current limit of 500 μA it follows that normal condition earth leakage currents must also not exceed this value. Higher earth leakage currents are only allowed in cases were interruption of protective earth connectors need not be assumed to be a single fault condition. This is the case for permanently installed devices and devices equipped with an additional (redundant) protective earth connector.

If during intended use the probability of contacting the patient either directly or indirectly via the user is negligible (and a related warning is included in the instructions for use) touch current limits may be exceeded at the following parts (§ 8.4.2 EN 60601-1):

- accessible contacts of connectors;
- accessible contacts of fuse holders;
- contacts of lampholders that become accessible during lamp exchange;
- parts behind covers of exchangeable components that can be accessed without a tool, or where users are instructed to use a tool (e.g. illuminated push-bottoms, indicator lamps, recorder pens, batteries or plug-in modules).

For parts accessible by a test finger, test pin or test cord, touch voltages and electric energies are limited to the following values (§ 8.4.2 EN 60601-1) in normal *and* single fault condition:

30 V AC (42.4 V_{peak}) or
60 V DC,

with the additional requirement that the energy shall not exceed 240 VA for longer than 60 s (14.4 kJ); released stored energy shall not exceed 20 J (at a potential difference up to 2 V).

Because of storage in capacitors energies and voltages could be accessible even after disconnection from the current source (§ 8.4.3, § 8.4.4 EN 60601-1). One second after disconnection or opening the accessible residual voltage shall not exceed 60 V DC between supply pins or between the enclosure and internal parts. It may be higher if the released charge is not larger than 45 μC. This requirement is particularly important for devices with large internal capacitors such as defibrillators, impulse lasers or X-ray generators.

10.3.2 Patient

Patients are protected by limiting patient leakage currents in normal and single fault condition (§ 8.7.4 EN 60601-1). Patient leakage currents for applied parts type B and BF are limited to ***100 μA*** AC; limits for applied parts type CF (for cardiac application) are reduced by a factor 10 to 10 μA AC (Table 10-1). Patient leakage DC currents of any applied part shall not exceed ***10 μA*** DC. In single fault condition every kind of leakage current (irrespective of type of applied part or time course) is allowed to increase up to ***5-fold***. In addition, if devices have more than one applied part, the overall patient leakage current with all applied parts connected together shall not exceed normal condition limits by more than 5-fold and single fault condition limits by more than 2-fold.

For patient auxiliary currents the same limits apply as for patient leakage currents.

Table 10-1: Limits for leakage currents in μA in normal condition

<table>
<tr><th rowspan="2">Current[a]</th><th colspan="2">Alternating current</th><th colspan="2">Direct current</th></tr>
<tr><th>Typ B, BF</th><th>Typ CF</th><th>Typ B, BF</th><th>Typ CF</th></tr>
<tr><td>Touch current</td><td colspan="2">100</td><td colspan="2" rowspan="3">10</td></tr>
<tr><td>Patient leakage current[b]</td><td rowspan="2">100</td><td rowspan="2">10</td></tr>
<tr><td>Patient auxiliary current</td></tr>
</table>

[a] In single fault condition all values are allowed to increase up to ***5-fold***.

[b] The overall patient leakage current with all applied parts connected together is allowed to increase up to ***2-fold***.

10.4 Failure assessment

Electromedical devices with failures and deficiencies are not rare. For new devices it can be requested that deficiencies be remedied prior to payment. However, decisions on how to proceed with deficient devices are more sensitive in the case of devices that are already in use. The reason is that repair and adaption not only cost money; in addition, the device might not be available for some time which may lead to consequential problems. Therefore, an important task is assessment of failures and deficiencies which have been encountered during recurrent testing in regard to their relevance for safety and essential performance. This is even more demanding because there are no rigid criteria. It is essential to put things into perspective and consider the context of a deficiency in regard to the kind of device, its inherent risk, relevance to the patient and availability of alternatives etc. Therefore, the same deficiency might be negligible in one case and need urgent remedy in another.

> **!** Failure relevance depends on the safety context

It has proven useful to classify failures according to the following scale:

Failure class 1 (tolerable failures): Tolerable failures include insignificant deficiencies such as a lost type label, however, provided it did not contain information which would require specific action or precaution. As an example, missing information on rated currents and input power could be tolerated if they were small enough and the device is not connected to a multiple socket outlet. However, if the nominal current would be higher than the rated current of conventional mains socket outlets (e. g. 16 A), this information would be relevant because of the risk of using unsuitable components, for example, when replacing the mains plug, with the consequence that the device could then be connected to an unsuitable mains socket outlet and consequently overload the installation. A dented metallic enclosure could be tolerated, however, only after checking for still maintained creepage distances, still reliable mechanical fixation of components and consideration of potential hygienic restrictions.

Failure class 2 (failures that allow delayed remedy): Class 2 failures are associated with non-acute hazards. They are not considered tolerable and need remedy, but would allow continuing operation for a limited time. It depends on the circumstances including available financial resources and manpower how quickly such failures should be corrected. As an example, a defective mains indicator lamp may be tolerated for a limited time provided switch positions are distinguishable by redundant means, a broken mains cable guard would not need immediate action in case of favourable concomitants such as good mechanical cable integrity and device placing and use which does not provoke excessive bending.

Failure class 3 (acute dangerous failures): Class 3 failures cause acute hazards and require immediate action; therefore, further operation of the device cannot be tolerated. Affected devices have immediately to be put out of service and their reuse reliably prevented. Whether repair is possible or the device would have to be discarded depends on the kind of failure, the age and condition of the device and the financial and personnel resources. As an example, a damaged mains cable with touchable bare leads, an interrupted protective earth conductor or defective alarms of an ECG monitor are not tolerable even for a limited time period.

Remark: Remedy of insulation failures by wrapping a band-aid around it is considered particularly dangerous since it simulates non-existent safety and, therefore, even enhances risk. Band-aids are no use as electrical insulation.

10.5 Documentation

Results of recurrent testing shall be recorded during the whole service life, from acquisition (receiving test) until the device is discarded (EN 62353). The initially measured values shall be recorded for comparison with subsequent measurements to allow identifying and assessing future degradation.

To be able to distinctively attribute test results to a particular individual device, units under test have to be clearly identified. Experience shows that devices don't remain in their original location; they might be borrowed and another individual device could have been brought back, they might be on repair and temporarily have been replaced by a hired one etc. Therefore, it is not sufficient to record the device type only. It is essential to clearly identify the particular device by its serial number or/and inventory number.

This is not only necessary at premises with several similar devices but clearly also in cases where only one single device of a certain type is used.

Test records shall contain the following (§ 6.1 EN 62353):

- identification of the testing body;
- tester's name;
- clear identification of the unit under test (e. g. company, type, serial number, inventory number);
- required accessories (as specified in the instructions for use);
- results of visual inspection;
- tests and measurements;
- measured values (with applied standard and measuring device);
- results of performance test(s);
- failure assessment (if applicable);
- final conclusion (failure classification);
- date;
- tester's (electronic) signature.

If a device is found to be acceptable and further operation is permitted, it should be marked by indicating the date of the ***next*** intended recurrent test (mm/yyyy). In case of faults assigned to failure class 3, the device should be put out of service and marked accordingly to prevent further use. The operator should be informed in writing of encountered failures and subsequent risks.

> *Remark: If permission is given, reuse of acutely dangerous devices could be reliably prevented by removing the plug of the mains cable.*

10.6 Visual inspection: Open the eyes!

It is well known that we are not able to see the world as it is. In fact, this was not even the objective of evolution. From all our senses, in particular from our eyes, an enormous information flood of more than 10^7 bit/s is continuously entering our brain. Conscious processing of such an amount of data would hopelessly overburden our brain. We are able to consciously perceive only a tiny portion, namely about 17 bit/s. This requires continuous and extreme data selection and filtering to extract only that part of the information that is considered relevant for the actual situation or our survival.

The data selection process is determined by congenital mechanisms and our perception habits. For this reason our experience, individual background and interests determine what kind of information we consciously perceive. As an example, someone who likes fashion will readily notice another person's clothing while car enthusiasts will hardly miss interesting car models even in dense traffic. Therefore, our perception process is selective. We consciously perceive and remember those items best that fit with our notions, while we tend to ignore or question things that challenge our opinion. This has been proven by investigations demonstrating e. g. that after watching the news we preferably remember information about the political party we prefer.

Viewing habits and experience also determine the result of visual safety inspections. Therefore, viewing and identifying device failures must be learned and trained.

> ! Viewing device failures must be learned and trained

As for visual safety inspections, our physiological restrictions mean that it is not sufficient just "to open the eyes"! Someone who does not know where to look will not be able to detect safety deficiencies. But even knowledge about safety standards and essential requirements alone would not be sufficient for reliable inspection. Similarly important is to follow a systematic procedure and have self-discipline. A tester whose attention confusedly jumps from one obvious deficiency to another will detect some, but may miss many others.

> ! Visual inspection requires knowledge, systematics and self-discipline

Depending on testing motivation objectives of visual inspection are different:

1. an intensive and accurate visual ***acceptance*** inspection, external and internal, is recommended when receiving devices without third-party approval (e. g. conformity class I and IIa). If safety deficiencies are overlooked at this time, the chance of (free) remedy by the manufacturer may be missed and the risk of later sorrows has increased.

   Remark: In the case of third-party certificates, visual inspection can be restricted to externally checking for potential damage during storage and transport.

> ! Acceptance tests prevent later sorrows

2. upon ***recurrent*** testing device records allow detection of whether device safety has been checked before. Therefore, attention is directed to degradation possibly caused through use, i. e. through stress, abrasion, ageing or contamination. Internal visual inspections are not necessary on a routine basis but are required upon indications for internal degradation such as a damaged enclosure, ingress of liquids, dust or dirt, excessive heating, soiled air filters etc.

Overview

Prior to testing it is important to clarify the device's intended purpose, safety concept, methodical risks and the existence of potential additional risk factors. This allows concentrating on critical aspects and properly assessing and classifying encountered deficiencies.

Medical systems should be particularly checked for exchanged, removed or amended components in comparison with the intended configuration and records of the previous recurrent test.

Special attention is required for critical safety aspects such as combinations of energy, high tension, operational sparks, movements, liquids, gases, pressure, heating, electromagnetic fields and radiation. Table 10-2 summarizes safety-relevant characteristics and subsequent required special attention to particular testing aspects.

Table 10 2: Device attributes requiring particular attention

Feature	Enhanced attention regarding
life-supporting function	function, alarms, batteries, accessories
emergency use	mechanical condition, protection from moisture, function, alarms, batteries, accessories
biosignal-monitoring	function, alarms, batteries, accessories
home-use	misuse damage, safe output values, accessories, understandable instructions for use
critical contact to patient	disinfection-related degradation (e. g. cracks, loss of elasticity), infection hazard
critical body region	disinfection-related degradation, infection hazard, patient leakage current, patient auxiliary current
extracorporeal blood circulation	function, protection from moisture, alarms, accessories, connectors
medical systems	components, overall connected power, leakage currents, protective earthing
critical use location	siting (explosion protection, electromagnetic interference)
mobile device	stability, lockability, overstressed mains connection
movable parts	stability, squeeze hazard, abrasion, emergency stop
use involves liquids	protection from spilled liquid (enclosure openings, inlet connector)
critical temperatures	isolation (discolouring, hardening, cracks), components
critical pressure	connections (leakage), alarms, accessories
critical gases	colour coding, connections (safety distances), flammable substances
critical measurements	calibration, alarms
critical substance delivery	dosage, protection from moisture, contamination, accessories, alarms
critical energy release	output values, accessories, alarms
critical radiation	radiation protection, protective accessories, alarms, key switches, door interlock
electromagnetic fields	siting, interference, alarms

10.6.1 Instructions for use

For the manufacturer apart from device markings the instructions for use are the most important tool to communicate with operators and users. It defines the intended use and performance, installation requirements and service life and it obliges users (e. g. by instructions, warnings and contraindications) and operators (e. g. by defining maintenance, recurrent testing and intervals) to cooperate to maintain safety and limiting liability. Operators are legally obliged to perform recurrent testing in an extent and in intervals as defined by the manufacturer.

10.6.2 Device markings

Device markings shall be durable and maintain readability during the whole service life. They must be positioned so as to allow users to read them from their intended position (§ 7.1.2, § 7.1.3 EN 60601-1). Movable devices may bear marks also on the sides and the back. Required markings are summarized in Table 10-3.

Table 10-3: Required marks on a device, if applicable

Item	Content
manufacturer	name and full address
identification	model and/or type
supply mains	rated voltage, current or power
power supply from accessory devices	voltage, phases, current and/or power (model or type of power supply unit)
applied part(s)	type
safety instructions	warnings, prohibition, precaution, commands text and/or symbols
protection	safety class, ingress protection
mode of operation	duty cycle (if applicable)
fuses	adjacent to (accessible) fuse-holders: voltage, current, characteristic
cooling conditions	e. g. water supply, air pressure (if applicable)
high tension	warning symbol

10.6.3 Device business card: Type label

It reflects not only politeness but also common sense to introduce oneself at a first meeting … or would you have confidence in someone who is completely unknown to you and whom you did not look at even once with more care?

Likewise, it makes sense and is important to acquaint yourself with a device before starting testing, for instance to clarify whether it is indeed a medical device, which critical performance and specific risks need to be considered, which safety concept was applied, and whether available supply (e. g. electric power, cooling media) is appropriate. For example, it could be that a UV-radiation device is not intended for medical use, a device originating from the USA or Japan (where mains voltage is about half as high as in Europe) could have an inappropriate default voltage setting, or a laser device with a rated input current above 16 A is not suitable for the available conventional 16-A socket outlet. In addition, there might be a label advising one to read the instructions for use to explain specific needs for installation, use and/or recurrent testing.

To provide users and technicians with comprehensive information, manufacturers are obliged to affix on their device a "device business card" in terms of a type label (§ 7.2 EN 60601-1). In most cases it is mounted on the rear of the device. It contains the most important information, mainly coded by numbers and symbols. To be able to understand their meaning, it is necessary to be acquainted with the most common symbols. Test yourself and try to figure out the meaning of the type label presented in Figure 10-1 and the symbols and signs contained in it. Afterwards try to link the information in the legend with associated symbols in the type label.

The design of the label is not standardized but left to the manufacturer. The condensed information of the example shown in Figure 10-1 means: The device is a laser unit, class 3B (indicated in the laser warning label), an electromedical device of conformity class IIb which was EC-type tested and produced under a quality management system (follows from the CE-mark, the elevated risk as concluded from the laser class 3B makes it conformity class IIb, the 4-digit identification number of a notified body indicates involvement of a third party – 0636 is assigned to the European notified body PMG, Graz University of Technology, Austria-, EC-type testing and quality management is demanded for conformity class IIb devices); the national safety mark of Germany (VDE) indicates conformity with all requirements of the applicable standards (rather than just the essential ones as confirmed by EC-type testing) and (some) market surveillance performed by the safety mark provider. The device is double insulated (safety class II) according to the associated symbol (two concentric squares); ex-

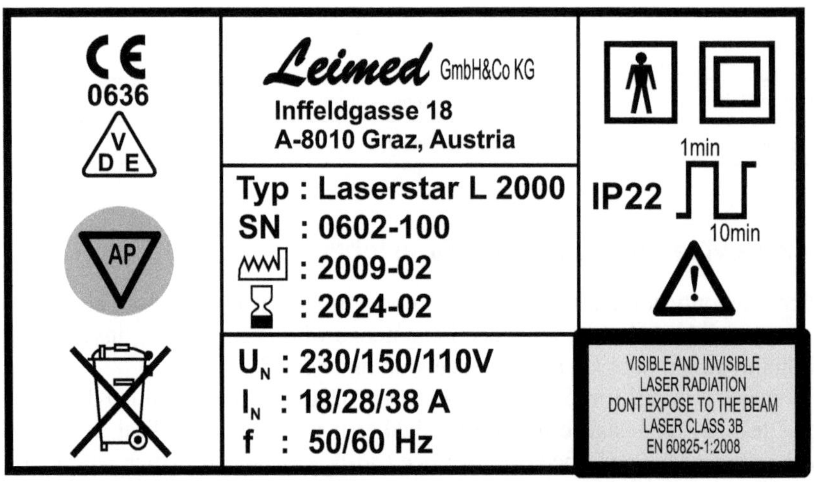

Figure 10-1: Type label of an electromedical laser device of laser class 3B, conformity class IIb, double insulated, intended for three different voltage levels, requiring a special mains plug, with a floating applied part, protected against touching live parts with the finger, splash-water proof, explosion protected against explosive mixtures with air, for short-term use, application is associated with increased risk, expected service life is 15 years, type-tested with third-party market surveillance; containing hazardous substances and not allowed to be disposed of as domestic waste.

plosion protection refers to mixtures of flammable gases with air as indicated by the symbol of the full circle with inscribed letters "AP". Protection against touching dangerous parts with the finger is indicated by the code IP2X and increased protection against ingress of liquids (splash-proof) as indicted by the code IPX2 (combined "IP22"). The device has a floating applied part (the symbol of the patient within the square) and is intended for short-term use with 1 min operation followed by a 10 min break. The expected service life time is limited to 15 years after production which follows from the expiry date indicated after the hourglass symbol and the year of production as indicated after the factory symbol. The device is intended for three different voltage levels according to the three given voltages separated by slashes (this requires checking whether the appropriate selection has been made). It has an elevated input current (18 A at 230 V). Therefore, it must be supplied by an electrical installation allowing rated currents higher than the conventional 16 A (following from the rated input current 18 A) and requires a mains plug different from the conventional 16 A-plugs. It must be checked whether former (inadequate) repair might have mounted an inappropriate plug and whether the supply circuit is indeed intended for higher rated currents. This requires checking the distribution box and verifying whether the overload circuit breaker is adequate. However, it is not uncommon that overload circuit breakers with elevated rated currents could have been installed without adapting circuit wiring. In combination with the selectable voltage levels and inadequate installation, increased fire hazard might arise from accidental 110 V setting (leading to 79-A input currents when supplied with 230 V mains); the resulting overload could lead to considerable fire hazard.

Elevated risks associated with the application of the device makes it imperative to follow carefully the instructions for use (as indicated by the triangular warning symbol). The device contains substances hazardous to the environment and must not be disposed of as domestic waste (which is indicated by the crossed out domestic waste symbol).

This example demonstrates that visual inspection without carefully reading (and understanding) the messages contained in the type label would miss important hazards and allow increased risks to persist. Besides this, the example shows that device testing might not be restricted to the device only but might need also further actions such as cross-checking the installation.

10.7 External visual inspection

The external visual inspection is an indispensable part of electromedical device recurrent safety testing. Because of the diversity of device design and appearance it is essential to stick to a systematic procedure and perform visual inspection step-by-step in a constant sequence: After general assessment of the device and its characteristic safety und functional aspects a glance follows the path of mains voltage from mains plug over the mains cable to the cable guard at the entrance point of the device, the fuses, the power switch and then passes to the controls, the patient connections and the accessories (Figure 10-2). This procedure can be summarized in the following ten test steps:

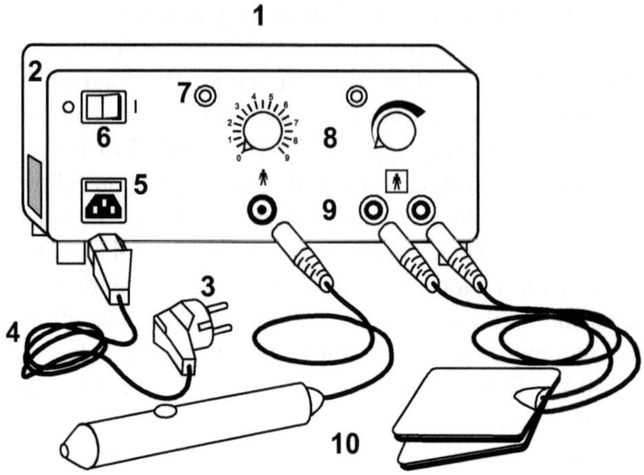

Figure 10-2: Test steps of external visual inspection

1. device in general (purpose, inherent risks, general aspects);
2. enclosure;
3. mains plug;
4. mains cable until device entrance point (cable guard and stress release);
5. fuses;
6. mains switch;
7. alarms;
8. controls;
9. connectors (applied parts, signal input/output, potential equalization);
10. applied parts and accessories.

1. Device

Before starting the detailed inspection, general aspects should be checked such as supply conditions (energy, gas, cooling media), placement, and environmental conditions. For that purpose it is necessary to know the device and its related requirements. Essential information can be found at the type label such as supply needs. The safety class can easily be determined only if all design rules have been met. Safety class II (double insulation) is indicated by the related symbol, battery devices by the lack of a mains connection. It is not required to mark safety class I devices – it can be assumed if other attributes are missing. Necessary but not sufficient clues for protective earthing are a metallic enclosure, a three-pin plug with a three-conductor mains cable and multiple mains fuses. In case of remaining doubts clarification could be gained by internal inspection.

> *Remark: Safety class II devices might also have a (double insulated) metallic enclosure, a three-pin plug and a three conductor cable – although in that case for functional earth connection only. Therefore, these attributes might not be sufficient to clearly identify the safety class. However, one glance into the inte-*

rior would clarify whether the earth conductor is connected to the enclosure (safety class I) or to an electronic board or metallic shield, and hence whether the earth connection is protective or functional.

- whether the existing ***electrical installation*** and the power supply circuit is sufficient or not can be decided once safety class and rated electric input values are known. Electric circuits of medical locations differ from general installations and must be equipped with a 30-mA residual current circuit breaker. The rated device input current allows deciding whether the actual mains socket outlet and the rated current of the overcurrent circuit breaker are sufficient.
- technical ***connection conditions*** for water, vacuum, compressed air and other gases are compared with specifications on the device and/or the instructions for use. Gas connections are checked for safe distances from electric connectors, correct colour coding and/or labelling and non-interchangeable safety thread (NIST) connectors.
- ***placement*** and installation of the device are checked in regard to
  - *cooling* conditions (are ventilator openings clear?);
  - *temperature* conditions (do close-by devices impair cooling, or contribute to heating, are there relevant heating elements and radiators, direct sunlight?);
  - *electromagnetic compatibility (*are sources of interference close-by such as ascending power cables, transformers, diathermy or RF surgery devices, or are devices vulnerable to electromagnetic interference such as biosignal recorders or monitors?).

  Remark: It must be taken into account that walls are no sufficient shield for against ELF magnetic fields, and appliances in the next room could contribute to interference and vice versa.

  - *explosion hazard* (is the device intended to operate in dangerous zones M or G such as a foot switch? Is it marked explosion-proof?);
  - *other potential critical environmental influences* (e.g. humidity, contamination).

General assessment of electromedical devices is based on EN 60601-1 which requires devices to be

- not ***overbalanced*** until an inclination of 5° (special attention needs to be given to devices built slim and high or having movable and protruding arms such as dental X-ray devices or patient lifters). Device stands and supports must be completely and reliably fixed; movable devices with castors and/or wheels must have locks or breaks (§ 9.4 EN 60601-1).
- ***robust***, devices shall withstand foreseeable mechanical stress during intended use (impacts, pressure, fall from a small height, horizontal movement against a 2 cm barrier) in particular, if intended for emergency or home use.
- ***mechanically safe***, devices shall not have dangerous corners or edges and be free of trapping zones. Danger of trapping and squeezing may particularly arise for devices with adjustable parts, motor-driven movements (e.g. hospital beds, height-adjustable devices such as a patient lifter) and rotating parts (e.g. centrifuges).

- ***stabile***, supporting or suspending parts for patients shall be designed for a load of 135 kg with a minimum safety factor of 2.5 (§ 9.8 EN 60601-1) except the manufacturer did mark the permitted load differently.

2. Enclosure

Enclosures are an important part of electric shock protection and must be provided during the whole expected service life. Protective enclosures must not be removable without tools. For plastic enclosures special attention needs to be given to cracks and mechanical weak parts such as ventilation grids, cooling slots, and to potential mechanical degradation due to chemical disinfectants. It should be checked whether flexibility is still sufficient and the surface has not become too rough. Metallic enclosure could exhibit deformations or dents. Signs of deformation, ingress of liquids, thermally induced discolouring, extensive dust, occlusion of ventilation filters are critical and require internal visual inspection to check for safety-relevant changes.

Openings must be checked in regard to protection against touching and ingress of liquids. Touch protection is insufficient if live parts (e.g. soldering, bare wires or circuits on electronic boards) are accessible across openings (e.g. by finger, test pin or test cord) or after removal of detachable parts (e.g. cover of an ECG paper roll or of a battery box) provided this is possible without a tool.

! Everything accessible without a tool is considered touchable

3. Mains plug

Any electromedical device shall have the means to simultaneously separate all poles from supply mains (§ 8.11.1 EN 60601-1). This can be done by a mains switch or a mains plug. Mains plugs are one of the parts most frequently exposed to mechanical stress. Therefore, they frequently exhibit deficiencies. If not sealed, plugs should be opened and connections and strain relief checked. If conductors are directly fixed by screws stranded conductors shall be protected from mechanical damage (Figure 10-5), for example, by end sleeves (to avoid the risk of interrupting single strings of stranded conductors with subsequent reduction of the cross-section and resulting excess heating). Screws must not clamp soldered stranded conductors either to avoid a dangerous positive-feedback process starting with deformation of solder, reduction of contact pressure, increase of contact impedance and excess heating which in turn accelerates solder deformation, impedance increase and heating which accelerates deformation etc. The protective earth conductor shall be lagging when connected, and the mains cable relieved from strain. The mains plug shall not be fitted with more than one power supply cable. If a plug connects DC voltages, dangerous reversions of polarity must be prevented (§ 8.2.2 EN 60601-1).

4. Power supply cable

Conductors of power supply cables shall have a minimum cross-section of 0.75 mm^2 (copper). There are only few exceptions defined in part 2 standards such as a permitted reduction to 0.5 mm^2 for safety class II nerve and muscle stimulators. At rated currents above 6 A the cross-section increases to 1 mm^2, above 10 A to 1.5 mm^2 and above 16 A until 25 A to 2.5 mm^2. The mains cable is inspected along its whole length for insulation damage and indications of excess bending (with the risk of wire breaks). Mains cables of movable devices (including hospital beds) are at particular mechanical risk. Therefore, it is recommended such devices be equipped with mains cable holders and helix cables. Moved mains cables with PVC insulation shall not be exposed to temperatures above 60°C. Therefore, particular attention should be given if cables could contact radiators or heaters. The protective earth conductor shall be an integral part of the mains cable and not be provided separately.

At the entrance point the mains cable must be protected from abrasion and excessive bending (§ 8.11.3.6 EN 60601-1). Its radius of curvature shall be not less than 1.5-fold the cable diameter. This could be reached for example by an insulating cable guard of sufficient length and stiffness, or by an adequately shaped opening. Excess bending protection is not considered necessary at permanently installed devices.

At the entrance point mains cables must not only be relieved from strain but also from stress and twisting. Screws, if any, that need be loosened when replacing the cable shall not simultaneously be used to fix any other component except the cable anchorage. Stress relief by knotting or by screws (metallic or not) bearing directly on the cable insulation is not permitted (§ 8.11.3 EN 60601-1).

Mains inlet connectors may be critical if enhanced protection against ingress of liquids is necessary such as for devices where accidental spilling of liquids must be assumed (e. g. nerve and muscle stimulators) or which require enforced protection against ingress of liquids. In such a case spilled liquid could enter interspaces, reach mains contacts, and make the dangerous mains voltage accessible.

5. *Fuses*

In contrast to household appliances electromedical devices must be protected against overload and short-circuit (§ 8.11.5 EN 60601-1) by fuses or overcurrent releases to keep consequences of a single fault as local as possible and avoid affecting other devices due to activation of the overcurrent circuit breaker in the distribution box. Devices with an earth conductor (protective or functional) shall have such means in each, and safety class II devices (without an earth conductor) at least in one supply conductor. It is recommended to place fuses before the mains switch to provide protection also in case of mains switch failure. External fuse holders must be designed so as to protect from touching active parts with a finger and during exchange of fuses (touch-protected fuse holders can be identified by their length which is about twice that of non-protected ones). If visual inspection did not find any or an insufficient number of fuses, the device might still meet the requirements because it is acceptable if some or even all fuses are placed inside the device. If there are doubts, internal visual inspection is necessary. Accessible fuses must be checked for touch protection and the intended rated values (voltage, current, blow characteristic). It is not uncommon that in case of blown fuses technicians don't have spare fuses with the demanded nominal values and provisionally insert higher-rated fuses and afterwards forget to change them again. Therefore, it is important that there be someone in the safety system that regularly checks fuses for appropriate values.

6. Mains switch

Any electromedical device shall have the means to simultaneously separate all poles from supply mains. This must not necessarily be a mains switch but if it is existent, it must meet all relevant requirements (e. g. switching all poles and providing at least 2-mm air clearance). In contrast to household appliances the mains switch of electromedical devices shall not be incorporated within a mains cable or any other flexible lead (§ 8.11.1 EN 60601-1). Apart from this, it is up to the manufacturer where he places the mains switch. However, switch positions shall be clearly visible and indicated by the standardized symbols “I” and “O” (see Table 10-4). To minimize human error, rocker switches must be mounted such that the “on” position is upward or to the right (in direct view of the switch). If an indicator lamp is foreseen it must be green (§ 7.8 EN 60601-1), see Table 9-4.

7. Alarms

Colours of indicator lamps are not freely selectable (§ 7.8 EN 60601-1). They shall meet the requirements for alarm signals (EN 60601-1-8). Alarms colour-coding and optical and acoustic time course are standardized to signal alarm priorities and urgency of required actions (see Chap. 9.2.2). Therefore, indicator lamps should be checked for colour coding. Red should be reserved for signalling acute danger only or for operation elements that need to be activated in such situations (e.g. emergency stop, § 7.8.2 EN 60601-1). Therefore, mains switch indicator lamps must not be red but have to be green.

8. Controls

Controls (e. g. switches, turning knobs) shall be indicated by figures, letters, symbols or other visual means and inform of the direction in which the magnitude of the related function changes (§ 7.4.2 EN 60601-1). If output values can reach dangerous levels, unintended changes shall be prevented (§ 12.4 EN 60601-1). This can be done by increasing awareness e. g. by requiring two independent actions such as safety covers (lift and select) or keypads (select and confirm). External visual inspection should concentrate on such preventive means, sufficient fixation of turning knobs, the agreement of the element position with the scaling, in particular the minimum and maximum position, and the function of the stop which prevents from overturning and unintentionally jumping from maximum to minimum position. In regard to usability, it should be checked whether control elements and functions could be confused. One means to reduce human error is selecting adequate symbols for switches. Table 10-4 summarizes the most important symbols.

Table 10-4: Symbols of switches

Symbol	Meaning
	mains switch "mains on"
	mains switch "mains out"
	mains push button: "mains on"/"mains out"
	push button: "standby"
	push button "in"
	"emergency stop"
	"device part in"
	"device part out"
	"device part standby"
	"function in"
	"function out"
	"function standby"

9. Connectors

All terminals must be clearly marked. This can be achieved by symbols or lettering (Table 10-5). The design of plugs and sockets must prevent from dangerous confusion. This applies in particular to applied parts. Particularly attention needs to be given to connectors of components of medical systems which were created by the user and where prevention of human error has not been part of the risk analysis. To avoid mistakes, device socket outlets must not fit within mains plugs, connectors and plugs of patient cables must nut allow unintended earthing (§ 8.5.2 EN 60601-1). If detaching connectors is dangerous, this must be prevented for example by locks (e. g. nerve and muscle stimulator, dialysis device, laser device). As a consequence, patient connectors must not be equipped with banana plugs, and devices shall not have sockets for banana plugs either.

> *Remark: The risk arising from confusing connectors is demonstrated from the following accident: When a mother visited her child, the nurse detached the ECG cables to allow the child to move. At the end of the visit, the mother wanted to reconnect the cables. She looked around for possible connection points and found a red coloured (emergency power supply) mains socket outlet right above the bed. Convinced that she had found the right connectors she plugged the cables in. Consequently an electric shock was delivered which killed her child.*

Table 10-5: Symbols for connectors

Symbol	**Meaning**
	applied part type BF
	potential equalization
	signal input
	signal output
	hand switch
	foot switch

At devices with connectors for flammable or oxidizing gases (e.g. O_2, N_2O) electric connectors shall keep a safety distance of at least 20 cm to them (e.g. medical supply units). Gas connectors should be uncoloured or exhibit the standardized gas-specific colour coding (ISO 32 /43/). Gas colours (Table 10-6) primarily signal the kind of hazard (flammable or explosive, toxic or corrosive, inert, oxidizing). Some frequently used gases are coded by specific colours. Connectors for mixed gases (e.g. air) may bear the colours of the gas components (e.g. black and white for compressed air connectors).

Remark: Particular risk arose from the adoption of the international colour code in the DACH countries Germany, Austria, Switzerland and Hungary. They had to change the colour of oxygen from the former blue to actual white and now face the risk of confusing oxygen with laughing gas whose colour was changed to actual blue. Lethal errors have already occurred.

Table 10-6: Colour coding of medical gases (ISO 32) /42/

Meaning	Colour
flammable or explosive	red
toxic or corrosive	yellow
inert	green
oxidizing	light blue
oxygen (O_2)	white
laughing gas (N_2O)	blue
nitrogen (N_2)	black
carbon dioxide (CO_2)	grey
compressed air ($O_2 + N_2$)	white/black

10. Applied parts and accessories

There are some changes in the definition of an applied part in the third edition of EN 60601-1 /27/ compared to the second edition. Now it is differentiated between applied parts, parts requiring similar protection to applied parts, and patient connections. In general, the term "applied part" is now restricted to those (conducting or non-conducting) parts of a device only that are intended to inevitably contact the patient during normal use (§ 3.8 EN 60601-1). Other not necessarily accessible parts with electric conducting connection to applied parts are now termed patient connections (§ 3.9 EN 60601-1). Therefore, depending on risk analysis, parts other than "applied parts" may require similar protection. As an example, the ECG electrode is considered an applied part, the connecting patient cable a patient connection and the electronic circuits inside the device until the separation from the mains part are named "other parts requiring protection similar to that of applied parts" (Figure 10-3).

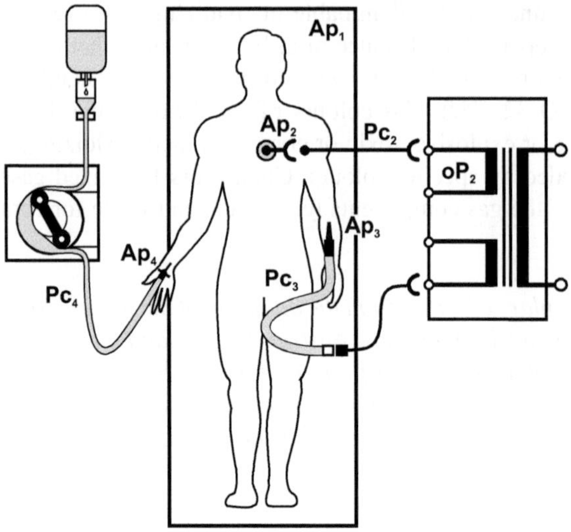

Figure 10-3: Examples demonstrating the differences between applied parts (Ap), patient connections (Pc) and other parts (oP), requiring a similar degree of protection. Ap_1 ... operating table surface (fabrics are not considered sufficiently insulating); Ap_2 ... ECG-electrode (including non-conducting adhesive surrounding and the plug); Pc_2 ... connection cable; oP_2 ... ECG-amplifier until separation from mains part; Ap_3 ... invasive blood pressure sensor; Pc_3 ... liquid column until pressure transducer; Ap_4 ... infusion cannula; Pc_4 ... liquid column until drop chamber

Visual inspection of applied parts and accessories should particularly concentrate on

- *labelling* (with symbols such as shown in Table 10-5). Applied parts connected to patient circuits shall be floating (type BF or CF) according to § 8.3 EN 60601-1. Symbols should be mounted at the patient connection points (except where this is impossible, in that case the applied part has to be marked (§ 7.1.10 EN 60601-1);
- *completeness* of required accessories;
- *suitability*: Applied parts and accessories (e.g. infusion sets) can be essential for safety. However, available products of various companies may be cheaper but not necessarily compatible with the actual device. As an example, the impedance of a RF surgery neutral electrode must fit with the electrode monitoring circuit of the device which is intended to detect electrode partial disconnection. Therefore, alternative products could critically delay alarms if they had an unsuitable impedance. As another example, infusion sets, in particular the inner dimension of tubes or the syringe for infusion pumps determine dosage accuracy, which may be compromised where unsuitable alternatives are used. Inspecting suitability includes also checking whether protective devices such as laser protective goggles are still suitable for the laser wavelength of the device actually used.
- *mechanical integrity*, in particular of connection cables (e.g. of RF surgery electrodes which frequently may be damaged by crock clip fixation in the operating theatre) or of handheld devices (e.g. applicators or hand switches).
- *no unintentional earthing* of patient circuits. To prevent this, contact pins of distal plugs of patient leads (e.g. ECG electrodes) must not be touchable and, if checked,

keep a distance of at least 0.5 mm from a plane surface (§ 8.5.2.3 EN 60601). Such a requirement would not be met by banana plugs.

- *expiry date* of accessories such as of self-adhesive electrodes which may cause dangerous electric current density increase or loss of function in case of (partial) detachment (e. g. defibrillator electrodes, ECG-monitoring electrodes, RF surgery neutral electrodes). Expired contact gel for defibrillator electrodes could cause uneven or unreliable contact which in turn might lead to burns or loss of function.
- *degradation and ageing* such as cracks of enclosures or loss of surface homogeneity, for example of reusable RF surgery neutral electrodes.

10.8 Internal visual inspection

While external visual inspection can and should be made also by the user, the inspection of the interior of a device is restricted to competent persons only. These are trained and aware of the potential hazards and consequences associated with the removal of protective enclosures such as damaging devices via electrostatic discharges. To assess air clearance distances it may be necessary to press against conductors to check their fixation or their potential displacement, or to gently pull components to decide whether they are still sufficiently connected. However, experience shows that in particular inexperienced testers tend to overdo such mechanical tests, and for instance rock a conductor so intensively and/or so long until initially sufficiently fixed conductors indeed are loosened. Therefore, it is necessary to caution against testing a device to death or leaving it worse than before. Therefore, as a rule, anything that can be decided visually should be done so and, consequently, prevented from unnecessary mechanical stress.

! After visual inspection devices should not be worse than before

Before starting with internal visual inspection it needs to be clarified whether at all devices must be designed to allow this. There are different options. On the one hand it is permitted to seal or cast devices, designing them as "fail-safe devices" without the need for inspecting the interior. In case of a failure these devices are intended to be just replaced by a new one. On the other hand, if maintenance or recurrent testing is necessary which might require opening the enclosure this must be possible without damage. The same applies if to maintain safety internal inspection is foreseen during the expected service life of a device. Particular attention is necessary if intended use is accompanied by additional risk factors such as movement, spilling of liquids, enhanced oxygen concentrations or the presence of flammable gases.

Although the variety in the external appearance of devices is already large this applies even more to their internal design. In addition, the packing density of components and electronic boards could be high and observation made more difficult. Therefore, internal visual inspection requires even more a systematic and constant approach.

First of all, it is recommended to get a general view on what parts are safety-relevant for users (such as the mains part) and for patients (such as output circuits). Then visual inspection starts, following the mains voltage from the entrance point and the mains cable anchorage to the mains terminal. Afterwards, if applicable, protective earthing and functional earth connections are checked. Then inspection continues to fuses and the wiring within the mains part up to the primary winding of the transformer and the separation of mains parts from secondary patient circuits. The transformer is the most important safety means and thus requires particular attention. Afterwards, separation of secondary parts and wiring from the mains part is inspected starting from the secondary winding of the transformer to the wiring of secondary circuits and its fixation up to applied parts and their terminals. Insulation is checked between parts of different voltage levels (e. g. mains voltage, electronic low-voltage level, and high-voltage). If possible, air and creepage distances are checked, in particular at electronic boards and connectors. Finally, used electronic components and critical regions are inspected where overstress (thermal or mechanical) or leakage could have occurred.

This strategy results in the ***ten-step approach*** of internal visual inspection (Figure 10-4), namely

1. power supply (cable anchorage to mains terminal)
2. earth connections (protective and functional earthing)
3. fuses
4. mains wiring (from mains terminal to separation)
5. mains transformer (including secondary fuses)
6. insulation (to enclosure, applied parts and between voltage levels)
7. secondary wiring
8. bare parts (air and creepage distances)
9. components
10. critical regions

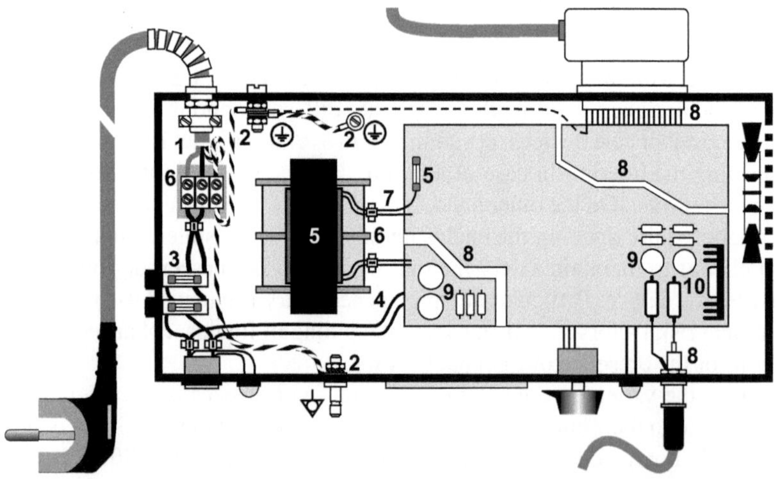

Figure 10-4: The ten-step approach of internal visual (1 … mains supply ⇨ 2 … earthing ⇨ 3 …fuses ⇨ 4…mains wiring ⇨ 5…mains transformer with secondary fuses ⇨ 6…insulation ⇨ 7 … secondary wiring ⇨ 8 … bare parts (air and creepage distances) ⇨ 9 … components ⇨ 10 … critical regions

1. Power supply

Mains cables are those components that are at most risk to be damaged. However, manufacturers are free to allow exchange of cables and to choose the way in which the cable is mounted to the device. There are three options available:

X-connection: The mains cable can be exchanged with conventional tools.

Y-connection: This allows the mains cable to be exchanged; however, the device requires knowledge on particular safety aspects (e.g. explosion-proof, water-proof, dust-proof). Therefore, mains cable exchange is restricted to specialists and, consequently, requires special tools such as triangular screw drivers.

Z-connection: This connection does not allow exchanging cables without damage (e.g. casted mains terminals or sealed enclosures).

Once the device is open the cable anchorage and strain relief can be inspected in more detail. As an example, strain relief by wire straps is not sufficient since it is not effective in relieving stress and twisting. If the mains cable is exchangeable, it should be checked whether this is easily possible without loosening internal connections or other mounted parts. Mains conductors should be connected at fixed terminals (usually a mains terminal block). Connection at other fixed connection points such as at EMC filters or internal overcurrent breakers is permitted in justified exceptional cases only. However, wire-to-wire connection is not acceptable.

If connections of stranded conductors are made by clamping, the turning screws must not expose them to mechanical tension. To achieve this, terminals could be equipped with metallic tongues (Figure 10-5). Screwless terminals are permitted if connection is possible without special preparation except for twisting of stranded conductors (soldering, cable sockets or cable eyes shall not be used). Protection of stranded conductors could also be achieved by using wire end sleeves.

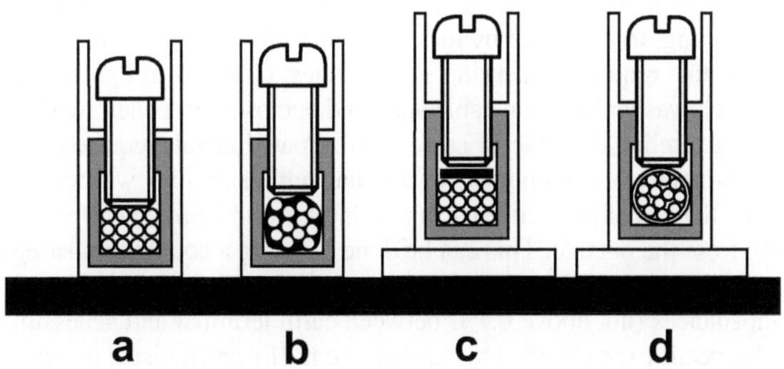

Figure 10-5: Unacceptable connections by screws directly acting upon stranded conductors (a) or upon soldered stranded conductors (b), and acceptable connections with mechanical stress relief by a metallic tongue (c) or wire end sleeve (d). Missing protection against unintended escape of an 8 mm wire at terminals a and b, protection against this by insulating supports at terminals c and d

Remark: Clamping soldered stranded conductors by screws is generally prohibited not only for medical devices. The reason is that contact pressure deforms solder which increases the contact resistor (R_c). In turn, contact temperature increases and due to its low melting temperature solder deformation progresses which in turn enhances resistor increase and contact heating (= $I^2.R_c$). This positive feedback continuously increases the contact resistor, heating and the risk of losing contact or/and causing fire.
Clamping of soldered stranded conductors is only allowed by contact springs which by their nature follow deformation and assure reliable contact.

! Clamped stranded conductors must not be soldered

Terminal blocks must be designed or insulated so as to prevent accidentally escaping single wires from contacting other conductors or touchable (grounded) metallic parts (e.g. device bottom). Frequently, it may be necessary to put a sufficiently overlaying insulating layer underneath the contact block (Figure 10-5).

2. Earth connections

When inspecting earth connections, two different cases have to be differentiated:

a) ***protective*** earth connections (safety class I): Protective earth conductors must have sufficiently large cross-sections to reliably carry short-circuit currents. Therefore, until the mains fuses (which limit short-circuiting currents) conductors must have the same cross-section as mains cable conductors. Smaller dimensions are allowed only in those areas where short-circuit currents are limited by internal fuses. Consequently, it is not permitted to reduce cross-sections of protective earth conductors by leading them partly across electronic boards along printed pathways. Protective earth terminals should be close to mains terminals and the conductor connected so as to assure that in case of mechanical strain it fails last. The terminal shall be marked by the protective earth symbol. Protective earth connections (including screws fastening metallic enclosures) should be mechanically protected from unintended loosening, for example by lock washers (Table 10-7). Contacting protective earth conductors requires observing several rules: Contacts to light metal require a hardened lock washer between cable shoe and enclosure to penetrate the oxide layer and assure a reliable contact (Figure 10-6). Coated metals require removal of the coating at protective earth contacts or alternatively using lock washers to penetrate the coating. Screws of protective earth terminals must be protected from unintended loosening from the outside. This can be done by using a counter nut safeguarded by a lock washer (Figure 10-6). The protective earth connections shall have sufficiently small impedances (not above 0.1 Ω between earth terminal and accessible metallic parts). This needs to be checked by measurement. However, visual inspection should pay attention to the earthing strategy. Multiple sequential contacts (with their multiple contributions to the overall impedance) should be avoided and a star point preferred. Attention should be given to indications of corrosion which would degrade contacts. Corrosion is supported by contact-voltages between two different metals. Therefore, nuts and screws for protective earth contacts should be of the same (non-corrosive) material. If earthing of enclosure parts is performed via device screws these connections like other protective earth connections should be mechanically safeguarded, for example by lock washers (Figure 10-6).

b) ***functional*** earth connections: They may be used in devices of safety class I or safety class II to ground metallic shields and so improve electromagnetic compatibility. However, in safety class II devices functional earth conductors must be insulated from accessible metallic parts similar to live conductors. Therefore, visual inspection should pay attention to wiring of functional earth conductors. Terminals should be marked with the appropriate symbol (Table 10-7). Functional earth conductors shall not be used for protection purposes to avoid compromising the concept of intrinsic safety.

Table 10-7: Symbols for earth connections

Symbol	Meaning
	Protective earth
	Functional earth
	Noise-suppressed functional earth
	Circuit ground

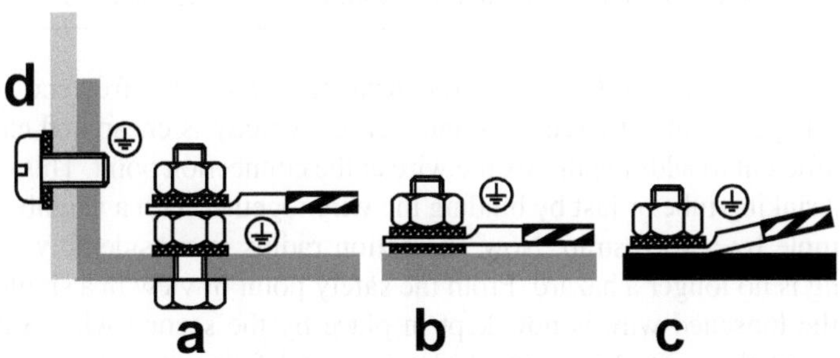

Figure 10-6: Requirements for protective earth terminals: a) Contact screw accessible from the outside with a lock washer and counter nut to protect from unintended loosening, and another lock washer to safeguard the contact; b) protective earth terminal not accessible from the outside with a lock washer to penetrate the oxide layer of light metal (e. g. aluminium) and another lock washer to safeguard the connection; c) protective earth terminal not accessible from the outside with a lock washer to safeguard the connection to metal; d) device screw to protectively earth the side panel with a lock washer to safeguard the connection.

3. Fuses

In contrast to household appliances electromedical devices must have an internal overload protection (e. g. mains fuses). This should ensure restricting breakdown to the device only which is affected by the single fault and avoiding interrupting operation of other devices in particular those that are life-saving or life-supporting. To avoid compromising this intention, the rated value of internal fuses must not be too high and should be chosen as low as necessary to reliably carry rated input currents. The number of required fuses depends on the safety class, in particular on the existence of a connection to earth:

- devices with earth conductors (safety class I devices or safety class II devices with a functional earth conductor) need to have fuses in all live conductors;
- for mains-supplied devices without earth conductor (safety class II) one fuse is sufficient;
- battery devices need a fuse only if a short-circuit could cause danger (e. g. fire). This could be neglected if the product of open-circuit voltage and short-circuit current is less that 15 W.

4. Mains wiring

Inspection of mains wiring should pay particular attention to the following aspects:

a) connection points: In principle, ***loosening of any wire*** at any connection point should always be assumed as a potential single fault, independent of the kind and place of connection. Be careful! This aspect is most frequently ignored by manufacturers and needs thorough checking. Therefore, in such cases it has to be considered which action radius is given to loosened wires and which consequences loose wires could have (e. g. bridging separation distances, causing short-circuits and connecting mains voltage to secondary parts or accessible metal parts).

! At any connection points loosening of wires must be considered

If loosening of wires is dangerous, for instance because the free end could bring mains voltage to patient circuits, in most cases remedy is cheap and easy. It is already sufficient to additionally fix the wire at the connection point. This can be done with a shrinking tube or just by binding the wire together with a neighbouring wire, for example by a wire strap. Now, the action radius is considerably reduced and loosening is no longer a hazard. From the safety point of view in a single fault condition, the loosened wire is now kept in place by the second wire. Simultaneous loosening of the second wire would be a second failure which according to the safety concept need not be considered (Chap. 9.2).

b) special precaution is recommended if a device contains ***different voltage levels*** such as 230 V mains and 5-V logic level and/or high-voltages such as in defibrillators or impulse lasers. It needs to be checked (if necessary with gentle pressure upon wires) to which extent wires could be displaced (during the whole expected service life). In the worst case it must be considered whether insulation or air clearance is still sufficient. However, it is important to apply the basic rule for safety considerations which says insulation that is insufficient for an actual working voltage needs to be considered non-existent and, consequently the so insulated conductors are considered as bare (therefore, double insulation would not be provided if one inadequately insulated conductor would be touching basic insulation).

! Inadequately insulated conductors are considered bare

If double insulation is required and conductors with non-equivalent insulation can touch each other, wires must either be reliably separated (e. g. by using wire straps) or insulation must be improved by an additional insulating tube.

c) wiring must be checked along its whole course in particular in regard to maintaining double insulation between mains and secondary parts. Special attention needs to be given to potential contacts to bare strip lines, to soldering at electronic boards and to bare component leads. To check potential displacements, it could again be necessary to gently press against wires.
d) if insulation tubes are used (e. g. to insulate bare soldered internal wire-to-wire connections or for providing additional insulation) they must be reliably prevented from safety-relevant displacement, for example by sufficient length, by fixing them with wire straps at both (!) ends or shrinking them along their entire length.

5. Mains transformer

A mains transformer is the most important safety-relevant component of an electromedical device. The task of a mains transformer is not only transforming mains voltage into the various required voltage levels but, most importantly, to reliably insulate secondary circuits from mains supply (and, if intended, also from earth potential such as for patient circuits type BF or CF). To achieve sufficient insulation several design options are available. Primary and secondary windings could either be situated in two separated non-conducting coil bobbins, or placed one upon the other separated by insulating interlayers (Figure 10-7).

a) it is frequently possible to identify insulation deficiencies of a transformer almost at first glance. The reason is that in most cases too little care is taken to consequently provide sufficient creepage distances. Basic insulation to the earthed core would require 4-mm creepage distance (see point 6 and 8); for double insulation between primary and secondary windings creepage distances should be 7 mm (for varnished wires the commonly required distance has been reduced by 1 mm). If one knows where to look, visual transformer inspection is easily possible. The most important check points for creepage distances are the inside corners to the transformer core and the separating layer between primary and secondary winding where the requirements frequently are not met (Figure 10-7). An alarming sign are brim fully winded coil bobbins. A frequent failure (although not visible from the outside) is insufficient separation of concentric windings by interlayers. The most common problem is that interlayers end right at the bobbin wall and are not extended to enlarge creepage distances (Figure 10-7).

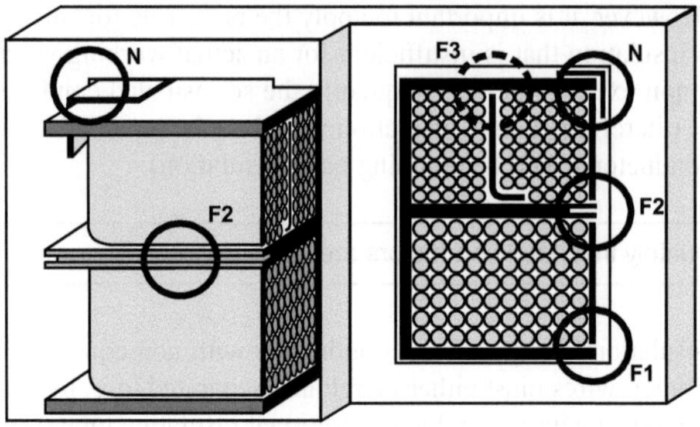

Figure 10-7: Safety-relevant checkpoints of a safety transformer
N … no deficiency: insulation elongates creepage distance to transformer core. F1 … failure: insufficient creepage distance to transformer core. F2 … failure: too narrow groove; therefore, insufficient creepage distance between windings. F3 … failure: insufficient creepage distance between windings (however, only visible by damaging the transformer)

Table 10-8: Symbols of transformers

Symbol	Meaning
	separation transformer requiring overload protection
	safety separation transformer requiring overload protection
	safety separation transformer, short-circuit proof
	safety separation transformer, fail-safe

b) Transformers should be protected against overload. This can be achieved by applying secondary fuses or by appropriate design. If fuses are foreseen it needs to be checked whether they are placed in every secondary circuit right before the first component (and not after rectifying circuits) and whether their rated values correspond with the values marked on the transformer type label (if existent).

c) The type of transformer can be marked by the appropriate symbol (Table 10-8). Safety separation transformers with double insulation between primary and secondary windings are marked by an escutcheon-like symbol. Open secondary contacts indicate that secondary overload protection is required, closed secondary contacts indicate short-circuit-proof design, and open secondary contacts with the letter "F" indicate that the transformer might fail in case of overload, but remain safe (fail-safe).

6. Insulation

Visual inspection of the insulation concentrates on sufficient dimensioning, reliability (e. g. of separation distances) and suitability of used materials (e. g. in terms of ageing, or resistance against moisture and inflammation), degradation, damage and thermal overexposure.

a) the required insulation strength can be summarized as follows (Figure 10-8):
The ***mains part*** must exhibit
– basic insulation to protectively earthed accessible metallic parts, and
– double insulation to floating accessible metallic parts, to applied parts, to signal input and to signal output parts.
Before fuses, other *active conductors* of the mains part require basic insulation between each, after them functional insulation is sufficient.
Live conductors and live parts require
– basic insulation to protectively earthed accessible metallic parts, and
– double insulation to floating accessible metallic parts.
Applied parts, patient connections and other parts deserving the same protection as applied parts (Figure 10-3) and require
– basic insulation to protectively earthed parts (except applied part type B) and to signal input- and output parts and;
– double insulation to accessible floating metallic parts.
Against each other conductors of applied parts shall exhibit at least functional insulation (Figure 10-8).

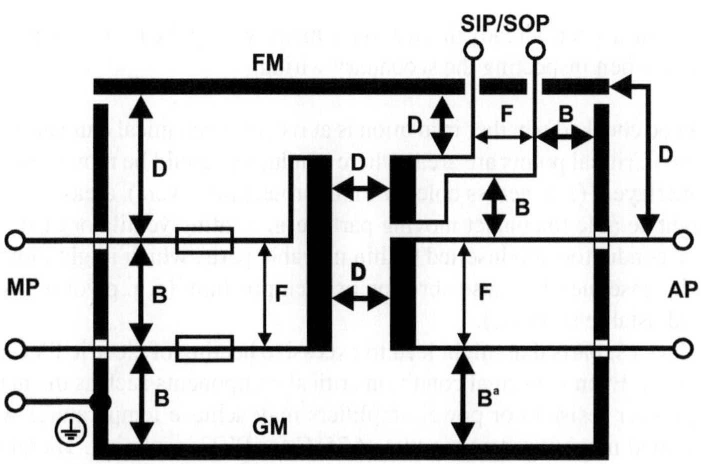

Figure 10-8: Required insulation between different parts of a device. MP … mains part, AP … applied part, SIP … signal input part, SOP … signal output part, B … basic insulation, D … double insulation, F … functional insulation, GM … grounded metallic part, FM … floating metallic part
[a] except applied parts type B (with direct contact to ground)

b) not all materials are suitable for electrical insulation. Insulation shall be durable, firm, moisture-resistant, flame-resistant and have a sufficient dielectric strength (§ 8.8.4 EN 60601-1).
   Therefore, ***not*** suitable for insulation are
   - wood and paper (flammable, not humidity-resistant);
   - deformable castings (not sufficiently rigid and stable);
   - coatings (not sufficiently durable and resistant against abrasion).

   Only of ***limited*** suitability are
   - PVC insulation (limited temperature resistance, looses flexibility and becomes cracked if heated above 75°C, at movable conductors above 60°C), applicable if prevented from overheating;
   - air (unreliable clearance distance), applicable if it is assured that clearance distances are kept for the whole expected service life (this may be checked by gently pushing against related parts).
   - Insulating foils (insufficient mechanical stability), applicable at a minimum thickness of 0.4 mm;
   - natural rubber (not age-resistant), applicable if intended for a limited time only together with instructions for preventive maintenance or adequate expiry date;
   - ceramic material (brittle, mechanical stability), applicable if tightly sintered and sufficiently protected it may be used for basic insulation;
   - electric tapes (insufficient reliability for the expected service life) applicable if sufficiently overlapping and additionally mechanically fixed; however, heat shrink insulation tubes should be preferred, anyway.

c) the required dielectric strength shall not be designed for the rated voltage but from the actual working voltage (in normal condition) that needs to be insulated. The minimum dielectric strength is 500 V.

7. Secondary wiring

In addition to the aspects already discussed for mains wiring the following points should be considered when inspecting the secondary wiring:

a) it should be checked whether insulation is at risk of mechanical damage, abrasion or overstress. Critical points are areas where conductors could be moved, bent or cross small interlayers (e. g. across holes without protection covers), areas where conductors might be able to contact moving parts (e. g. rotating ventilator blades, motors) or where conductors are inserted within movable parts, which might move conductors and, consequently, cause abrasion or even pinching (e. g. pivot or rotary arms, height-adjustable supports).

b) within devices, parts that might lead to excessive heating of electric PVC insulation are frequent. Even in normal condition critical components such as the mains transformer, power resistors or power amplifiers may achieve temperatures well above the permitted maximum temperature of 75°C for PVC insulation. Therefore, PVC-insulated conductors must be kept away from such components. Even higher temperatures could be encountered within devices containing heating elements (e. g. infrared radiators, patient warmers, thermocautery devices). Wiring must be checked for indications of excess heating such as discolouration or loss of flexibility.

c) colour coding of internal wiring can be chosen without restriction. The only exceptions are earth conductors independent of their special function, whether it might be protective, functional or potential equalizing. All earth conductors should be insulated yellow-green (with at least 30% proportion for either yellow or green). In exceptional cases where this is not possible, they should be marked yellow-green at least at the terminals. Such exceptions are connections between device parts with multi-conductor cables where using a single cable would not provide a sufficiently large cross-section. Therefore, in this case connecting several cables together for a protective earth connection is permitted. Another exception would be a large cross-section bare copper netting band to connect parts of devices with expected large short-circuit currents.

8. Bare parts (air and creepage distances)

Bare live parts must be insulated by air and creepage distances. For instance, this applies to electronic boards, soldering points, component wires and connection points of switches, connectors and plugs. Where gaps and distances are reliably filled with insulating compounds, air and creepage distances do not exist.

a) ***air distances*** are the shortest possible distances across air, across gaps and across uncemented insulating barriers (Figure 10-9). If the air distance is interrupted by floating conducting parts distances below 1 mm are not considered. If gaps are larger than 1 mm no creepage distance exists, and the separation distance is measured (and assessed) as air distance only.
b) ***creepage distances*** are the shortest possible distances along surfaces. Grooves smaller than 1 mm are ignored. Barriers placed on the surface are enlarging creepage distances only if they are affixed such as preventing dust and moisture penetrating the gap. Uncemented barriers are ignored. Measurements are made across them. The reason is that such gaps do not prevent electric currents from flowing. In contrary, due to capillary effects moisture could be attracted and gap conductivity enhanced. If the creepage distance is interrupted by floating conducting parts distances below 1 mm are not considered. Prior to measurement screw heads and nuts are brought in the most unfavourable position, grooves smaller than 1 mm are ignored (Figure 10-10).

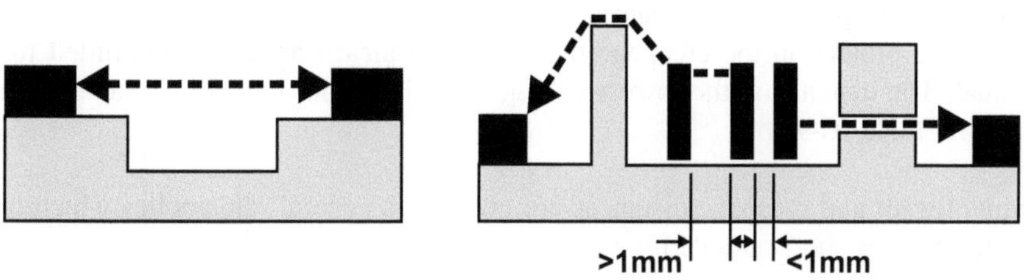

Figure 10-9: Measurement of air distances across gaps (left) and around insulating barriers and across uncemented insulating barriers (right). The distance between the two conducting parts <1 mm is not considered.

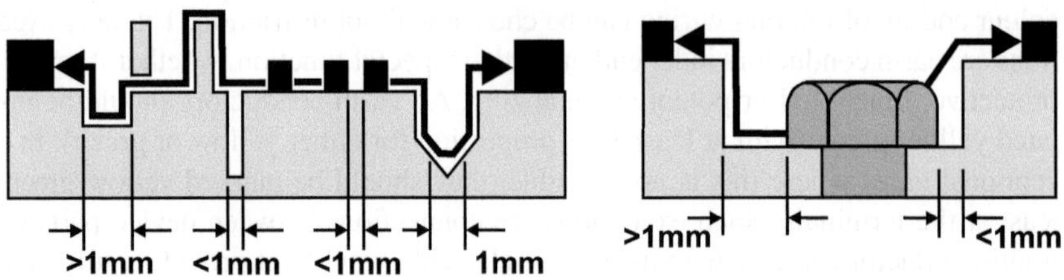

Figure 10-10: Measurement of creepage distances along grooves >1 mm, across uncemented barriers, along gap-free barriers, across small grooves <1 mm with distances between conducting parts <1 mm ignored (left), and with screw heads in the most unfavourable position (right)

When checking distances special attention should be given to common pitfalls such as:

- electronic boards with markings and inscriptions of conducting copper: This could lead to critical short-circuiting of initially sufficiently dimensioned creepage distances;
- mounting screws and metallic washers could critically reduce creepage distances;
- frequently distances between soldering eyes are a problem (e.g. at multipoint connectors). If they are less than 1 mm they would be considered short-circuited (except for functional insulation). Therefore, problems could be encountered even at seemingly large distances of the connected conductors;
- late corrections of the electronic board layout that are made by free wiring could cause safety problems if wires were not prevented from loosening under single fault condition and could bridge creepage distances.

c) to find the required ***limits for air and creepage distances***, first it is necessary to identify whether distances should protect users or patients or whether they aim at maintaining function only. Apart from this aspect, distances are determined for the working voltage they need to insulate. Tables with air and creepage distances can be found in EN 60601-1 /27/. Double insulation requires double distances. For creepage distances along inorganic insulation material (e.g. glass or mica) with similar properties to creepage paths in ceramics, creepage distances can be as small as air distances (§ 8.9 EN 60601-1).
For patient protection the minimum value (even at lowest working voltages) for air distances is 0.8 mm (for 230-V mains voltage 2.5 mm is required). Creepage distances are larger than air distances.
For user protection the creepage distances and air distances are permitted to be equal. The distance at the lowest voltage level is 1 mm, for 230-V mains voltage 2 mm (Table 10-9).

If limits for air and creepage distances are not met, a general rule applies which is to consider them as short-circuited.

! | Insufficient air and creepage distances are considered as short-circuited

Table 10-9: Minimum air and creepage distances (in mm) in dependence of the working voltage UB (in Volts) for the lowest voltage level and mains voltage in case of low pollution /27/.
d_a … air distance, d_c … creepage distance

Working voltage		Patient safety		User safety	Function	
DC	AC	d_a	d_c	$d_a = d_c$	d_a	d_c
≤17	≤12	0.8	1.7	1	0.4	0.8
354	250	2.5	4	2	1.6	3

Conditions that need careful clarification by visual inspection are tracks on electronic boards that lead close to the edges of the board. If boards are mounted with small air gaps, air and/or creepage distances to the metallic enclosure could be too small. Other examples of potential problems with creepage distances are electronic boards mounted with the soldered side close to the metallic enclosure, easily deformable bare wires or soldering tags, creepage distances between transformer windings and from windings to the transformer core.

9. Components

When visually inspecting components, attention should be given to rated values in regard to compliance with actual operation conditions and mechanical aspects such as reliable mounting of large capacitors, connections and batteries.

a) ***compliance*** of components with use conditions is checked by comparing their nominal values with rated values of the device (e.g. rated input current, mains voltage, operation characteristics and/or permissible temperature). Non-compliance is found frequently at mains switches and safety transformers. It may be that mains part components may not be designed for rated input currents, other components such as electrolytic capacitors may not be intended for the given ambient temperatures, or mains switches may break a single pole only.
b) ***mechanical aspects*** comprise fixation of components, stability (problems can be encountered at heavy components such as large electrolytic capacitors which are frequently insufficiently mounted by their connecting wires only, but should be additionally fixed by for example hot gluing). Plugs and jacks should withstand rough handling during connection and disconnection and should be checked for damage.
c) ***assembling*** aspects such as placement of critical components (power resistors, power amplifiers, transformers. ventilators, gas tubes and connectors) relative to each other, to the enclosure and to the wiring.
d) ***connections*** of conductors at components, in particular transformers, plugs, output controls etc., might violate creepage distances because of basic design problems and/or high packing densities, but also because of inadequate soldering. Therefore, creepage and air distances among contacts and to accessible parts need careful checking.
e) ***batteries*** need to be inspected for example for integrity, tightness and/or age and whether there are indications for necessary replacement, in particular in devices with critical function such as defibrillators, ambulatory infusion pumps, devices with buffer batteries for alarm functions, or battery-supplied devices.

10. Critical regions

Internal visual inspection is concluded by checking critical regions such as places with excess heating, high mechanical stress (e. g. joints, supports, handles and grips), excessive pollution for example of ventilation filters, due to DC motors with abrasion from contact brushes (e. g. centrifuges), paper fluff within ECG recorders or cathode ray tubes with electrostatically attracted dirt), liquid containers and battery boxes with potentially leaked conducting or aggressive liquids etc.

10.9 Options for corrections

In practice, meeting required distances is not self-evident, to the contrary. However, while insufficient air distances usually can be easily managed by inserting insulating layers, correcting insufficient creepage distances requires more effort. Too short creepage distances can frequently be found on electronic boards between mains and secondary conductors, at connector sockets, or at (grounded) mounting screws. Printed metallic markings may also compromise clearance distances.

Since pure listing of deficiencies does not already solve the problem, the question arises how such deficiencies could be corrected. If encountered during acceptance testing the situation would be more comfortable for the tester. Complaints could simply be passed on to the manufacturer or distributer and payment could be made dependent on adequate remedy of deficiencies. However, with devices already in use (and paid for) the problem frequently stays with the operator. On the other hand, manufacturers who already have a number of deficient electronic boards in stock could also seek ways to correct the faults.

Assessment

First of all, it is necessary to clarify the safety relevance of deficient creepage distances. It is evident that a deficiency may reduce reliability of the device. However, whether it may lead to an unacceptable risk depends on further circumstances. To assess safety relevance it is tested, what the consequences would be from short-circuiting through deficient distances. It must be kept in mind that this would not be considered a single fault. Since the safety objective is meeting leakage current limits, this could be clarified by measuring resulting earth leakage currents, touch currents and/or patient leakage currents. If normal condition limits are met, reduced creepage distances could be accepted without further action. However, prior to short-circuiting it should be clarified that the device or components would not be damaged by the short-circuiting.

! Short-circuiting a deficient creepage distance is not considered a single fault

Reduced creepage distances could be tolerated in the following cases:

- between opposite polarities within the mains part; however, only after the fuses, in which case short-circuiting would not be dangerous but only cause blown fuses.
- if creepage distances provide basic insulation (e. g. between applied part and the enclosure or between live parts and the enclosure) if short-circuiting does not cause patient leakage currents and touch currents exceeding their limits for normal conditions.

With the above-mentioned checks no further action would be required. However, even if this would not be the case, insufficient creepage distances need not be considered an unsolvable problem.

Remedy

Of course, redesign of electronic boards would be the best option, albeit an expensive solution. But there could already be a considerable number of boards in stock and their disposal could lead to a significant financial loss. Therefore, there are still other alternatives to consider:

Terminal block

- frequently, problems may exist with wires escaping at (mains) terminal blocks (remember, that an 8-mm free wire must not contact the protectively earthed enclosure). In that case it is already sufficient to mount an additional insulating layer of at least 0.4 mm thickness underneath the block if at all sides it extends sufficiently beyond it.
- if escaping wires could cause dangerous contacts with adjacent conductors, the use of a longer terminal block could allow leaving a spare terminal free and thus enlarging the distance between the critical clamp points. If this is not possible due to restricted space, an insulating plate could be tightly glued in between the two, for example using hot glue.

Electronic boards

For electronic boards deficient creepage distances could be managed in different ways:

- the easiest, although an unattractive solution, is to cover critical regions with an adhesive non-thermoplastic casting compound (varnishing would not be sufficient and silicone would not be adequate because of its insufficient adhesion and mechanical stability).
- frequently, deficient creepage distances are found in few places only for example at edges where a conductor critically approaches another circuit. In these few cases, mechanically removing the copper layer and rounding of edges could be sufficient to enlarge the distance.
- if the creepage distance is not smaller than the required air clearance, another solution could be to mill a gap of at least 1 mm width into the electronic board. In that

case the entire distance is considered an air distance which is usually only about half the required creepage length. Of course, care must be taken not to critically compromise the mechanical properties of the electronic board.

- if creepage distances are deficient along a larger pathway and for mechanical reasons milling a gap would not be acceptable, another option could be to interrupt this pathway along a sufficient length and replace it by an external conductor. Of course, this conductor must be additionally fixed at both terminal points (e. g. by hot glue) to prevent it from bridging critical separation distances or contacting other parts in case of loosening (single fault condition). In addition, it might be necessary to provide double insulation of the external conductor to other bare soldering points, for example by covering it with an insulation tube.

Transformers

Deficient creepage distances at or within transformers could be corrected, depending on the site of their appearance, by introducing additional insulation layers (e. g. around the corners of the coils). Manufactures may have also the option to cast the entire transformer. With this measure creepage distances would no longer exist and remaining problems could only be expected at wire outlets to external contacts.

10.10 Measurement

10.10.1 Safety parameters

To assess a device's safety it is important to measure safety parameters such as leakage currents and, in case of doubts, insulation impedance (EN 62353 /33/). In addition, the protective earth impedance of protectively earthed devices must be measured.

Measurement results of new devices should be recorded for comparison with subsequent recurrent tests. However, these ***reference values*** must be updated if the device or its configuration has been changed, for example after repair or after changing components of a medical system.

Medical systems connected to the mains by one multiple socket outlet are treated like one single device. However, if components are supplied via an own mains cable which is not fixed to the system they are tested both as individual devices and together with the entire system (§ 4.1 EN 62353).

Remark: a medical system is considered any combination of devices connected together via a functional connection and/or via a multiple socket outlet.

10.10.1.1 Protective earth impedance

Only those accessible metallic parts of safety class I devices must be protectively earthed that can become live in single fault condition. To provide sufficient protection, the impedance of such connections must be sufficiently low. Already the term "impedance" (instead of resistance) indicates that in case of doubts measurement must be performed with AC currents. As an example, the impedance of a mains cable rolled up in a cable reel could exceed limits while in the unrolled position or measured with DC currents limits could be met. Therefore, the worst-case measurement with AC currents must be recorded. The impedance limits are (§ 8.6.4 EN 60601-1):

0.1 Ω for ***internal protective earth connections*** (measured between protective earth terminal and accessible parts). For devices already in use ageing is considered by increasing the limit to **0.2 Ω** (§ 5.3.2.2 EN 62353);
0.1 Ω between both terminals of a mains cable (without relaxation for ageing);
0.5 Ω for a medical system already in use (measured between the mains plug of the multiple socket outlet and accessible metallic parts of system components).

Therefore, between the protective earth terminal of the mains plug and accessible metallic parts of the enclosure the following limits for the protective earth impedance must be met:

0.2 Ω at new devices (acceptance test);
0.3 Ω at devices already in use (recurrent test);
0.5 Ω at medical systems supplied via a multiple socket outlet.

Measurement is performed with current amplitudes of at least 200 mA and a free running voltage of less than or equal to 24 V. However, measurement to one single arbitrarily selected point is not sufficient. Reliable testing requires checking all relevant (separated) touchable metallic parts and recording the maximum value.

In general, for impedance measurement the device has to be separated from the mains (and protective earth). However, permanently connected devices may be checked without separation from the mains and protective earth. In that case the impedance is measured parallel to the mains connection until a neighbouring earth contact (e. g. at a mains socket outlet or a potential equalization point). Additional impedances of contacts within the mains supply may be considered by subtracting from measured values (§ 5.3.2 EN 52353).

Remark: DC impedance measurements are permitted provided they are repeated with reversed polarity of the mains plug. If relevant inductances are expected (e. g. in the case of cable reels) DC measurement could lead to underestimation.

If limits are exceeded, a knee-jerk reaction should be avoided. This must not already be interpreted as non-compliance. The reason is that excess could be tolerated under some circumstances (which have to be checked prior to a final conclusion) such as

1. if parts cannot become live in single fault condition. However, an internal inspection would be necessary to clarify this (see Sect. 10.8).
2. if the safety objection (meeting leakage currents) remains met in spite of short-circuiting basic insulation to the suspected accessible part (§ 8.6.4 EN 60601-1). This could be the case if the power supply is earth-free or currents in relevant circuits are limited and hence critical currents could not be provided.

10.10.1.2 Leakage currents

As a general rule, leakage current measurements must be performed together with the intended accessories. Components of medical systems are tested as individual devices if they have an own mains cable that is not fixed to the system or its multiple socket outlet. For measurement, three options are available:

- ***direct measurement*** of equipment leakage currents similar to EN 60601-1, with the device connected to the mains and operated in standby.
- ***simplified measurement*** of alternative leakage currents according to EN 62353 with both input and output short-circuited. In that case the device cannot be operated.
- ***differential measurement*** of permanently installed devices according to EN 62353.

Remark: It may be necessary to separately measure leakage currents to not protectively earthed (double insulated) metallic parts of safety class I devices since lower leakage current limits are set for these parts.

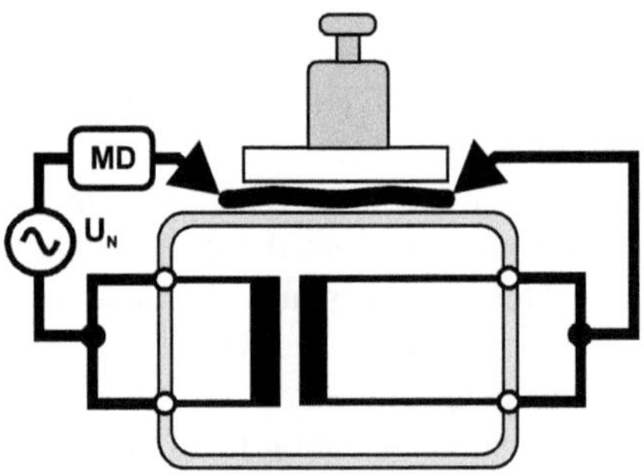

Figure 10-11: Contacting non-conductive enclosures by an aluminium foil pressed upon it by a symbolic load (the example shows measurement of the alternative device leakage current)

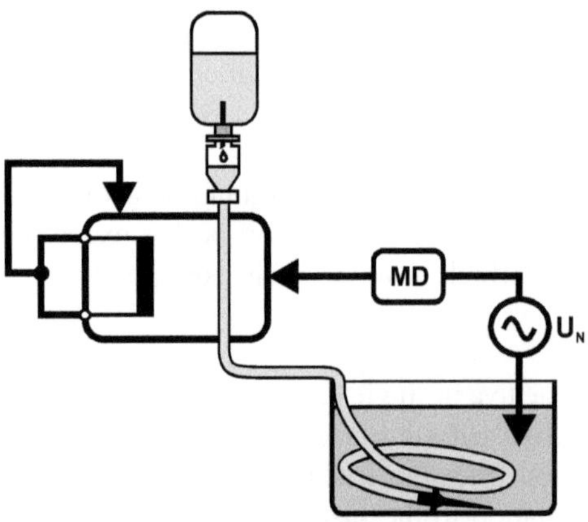

Figure 10-12: Contacting difficult to contact applied parts by immersion in a saline solution (the example shows measurement of the alternative patient leakage current of an infusion pump).

If the enclosure or parts of it consist of insulating material they are contacted by a metallic foil (e. g. a domestic aluminium foil) of a size matching the kind of expected contact (e. g. 20 × 10 cm for hand contact). To allow contact with the entire foil it usually needs to bear a load (Figure 10-11).

In case of difficult to contact applied parts such as infusion sets or catheters contacting can be performed by immersion into a container filled with physiologic saline solution (Figure 10-12).

1. Enclosure leakage current

The device leakage current is that current that flows between accessible parts of the enclosure across a person to earth during standby operation.

Direct measurement

For recurrent testing direct measurement according to the generic device standard has been simplified by the standard EN 62353. Although the device is still connected to mains voltage and ready to operate, currents are not measured at 110% mains voltage but at actual mains voltage; besides this, the various single fault conditions foreseen in the generic standard are not checked. Protective earth is disconnected and applied parts are short-circuited and connected to the enclosure (Figure 10-13). The leakage current is measured with the measuring circuit simulating a patient's body impedance and excitability (Figure 8-20). Single-phase devices are measured with either mains plug position.

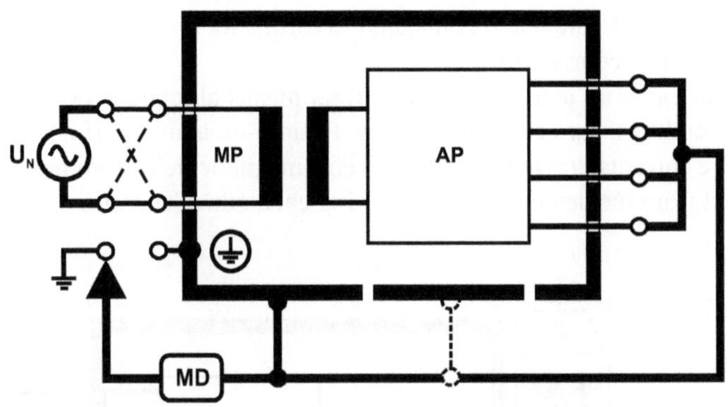

Figure 10-13: Direct measurement of the enclosure leakage current (the dotted line indicates another measurement position). MP … mains part, AP … applied part, MD … measuring device, U_N … mains voltage

Alternative enclosure leakage current

The measurement of the alternative enclosure leakage current is performed with a simplified measurement circuit with short-circuited mains input part. Patient connections are also short-circuited and connected to the enclosure. The measurement is performed with the measurement device (Figure 8-20); mains voltage is applied between short-

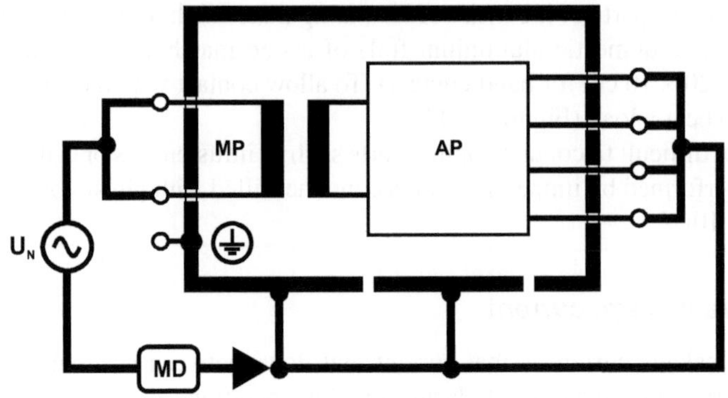

Figure 10-14: Measurement of the alternative enclosure leakage current. MP … mains part, AP … applied part, MD … measuring device, U_N … mains voltage

circuited mains part and enclosure (Figure 10-14). With this approach, the entire insulation between mains part and all remaining accessible parts (including applied parts) is checked.

Differential measurement

The differential measurement is based on the principle of the residual current circuit breaker. It is especially useful in case of permanently installed powerful devices (e. g. X-ray generators). The device under test remains connected to the mains, applied parts, if existent, are short-circuited and connected to earth. Not protectively earthed metallic parts are also connected to earth.

For measurement, all active conductors (that means all conductors except the protective earth conductor) are encompassed by a clamp-on ammeter (Figure 10-15). The sum of leakage currents flowing to earth prevents magnetic fields (produced by currents flowing to and from the device) from cancelling each other. Therefore, a residual mag-

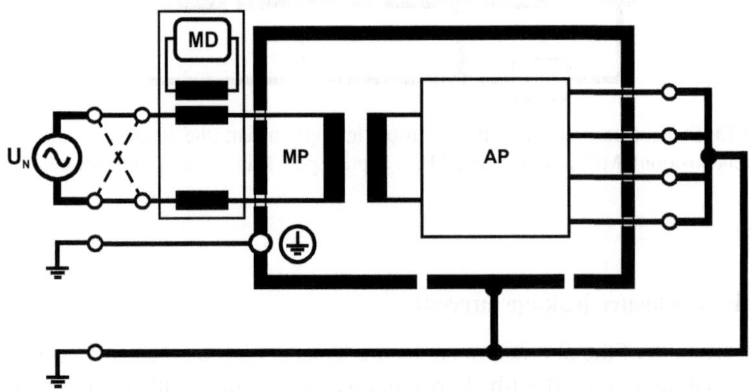

Figure 10-15: Differential measurement of the entire equipment leakage current. MP … mains part, AP … applied part, MD … measuring device, U_N … mains voltage

netic field induces a voltage into the clamp which is proportional to the current difference and hence to the entire equipment leakage current. Its frequency assessment is performed as usual by the measurement circuit. However, since clamp-on ammeters are not available for small currents, small leakage currents may not be measurable by this approach.

2. Patient leakage current

The patient leakage current is flowing from the applied part across the patient to earth. If not otherwise defined by the manufacturer it is measured at floating applied parts (type BF or type CF) only. Patient leakage currents of applied parts are measured separately with all terminals of an applied part short-circuited while terminals of other applied parts remain open. The highest measured value is recorded.

Direct measurement

Measurements are performed with the device connected to the mains. At devices of safety class I the protective earth connector is not interrupted, floating accessible metallic parts are connected with earthed parts (Figure 10-16). Non-conducting enclosures of devices of safety class II are contacted with a metallic foil (Figure 10-11).

The leakage current is measured with the measuring circuit simulating a patient's body impedance and excitability (Figure 8-20) while mains voltage is applied between the short-circuited applied part and earth. Single-phase devices are measured with either mains plug position.

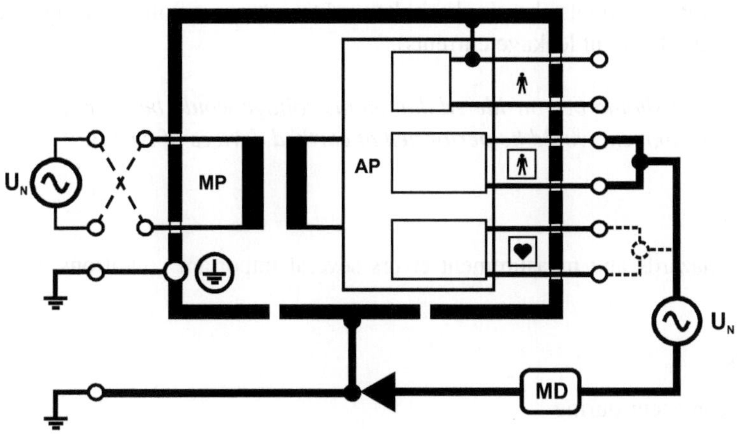

Figure 10-16: Direct measurement of the alternative patient leakage current. MP … mains part, AP … applied part, MD … measuring device, U_N … mains voltage

Alternative patient leakage current

The measurement circuit for the alternative patient leakage current essentially mirrors the simplified measurement circuit for device leakage currents. Now the mains input part is short-circuited and connected to the enclosure. The mains voltage is applied

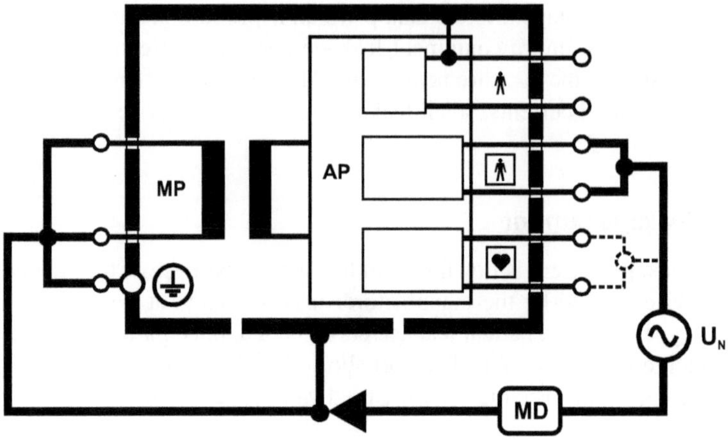

Figure 10-17: Measurement of the alternative patient leakage current (the dotted lines indicate another measurement). MP … mains part, AP … applied part, MD … measuring device, U_N … mains voltage

between the short-circuited applied part and the enclosure (Figure 10-17); the remaining patient terminals remain open. The measurement is performed with the measurement device (Figure 8-20). With this approach, the entire insulation between one applied part and accessible parts and the mains part is checked. Measurements are performed separately for each applied part type F. The highest measured value is recorded.

This situation equals the single fault condition "mains voltage at applied parts" (with increased patient leakage currents).

Remark: It should be considered that mains voltage would be short-circuited if this measurement would be performed at earthed devices of type B.

3. Pitfalls

To avoid hazards and measurement errors several important conditions must be observed.

Hazards

Hazards can occur during

- *alternative measurements* if an earth-related mains voltage is applied instead of test voltages generated by an insulation transformer. The reason is, as measurement circuits show, that one pole of mains voltage is connected to the device enclosure. Therefore, the dangerous voltage might become accessible.
- *direct measurement* of the enclosure leakage current. The reason is that the protective earth conductor is disconnected during measurement and connection to earth performed via the measurement device which includes an impedance of ≥ 1kΩ. Therefore, safety in single fault condition is compromised.

- *direct and alternative measurements* of patient leakage currents of earthed applied parts (type B). The reason is that the measurement voltage is short-circuited by the existing earth connections (irrespective of whether it is earthed or not).

Measurement errors

Leakage current measurements can suffer from measurement errors for different reasons. However, they all have one thing in common: All of them lead to reduced measured values and, hence, may probably mask non-existing safety. Therefore, if recurrent leakage current measurements result in unusually low values – don't be happy about a seemingly excellent device status but consider instead what could have gone wrong with the measurement.

! In most cases unusually low leakage currents result from measurement errors

Unfortunately there are several possibilities for such a slip up:

- a frequent cause is *unintentional earthing*. In the case of an earthed voltage supply or measurement voltages, leakage currents use all existing options to flow back to earth but prefer the route with lowest impedance. If devices are not floating, only a small part of the leakage current is flowing across the 1 kΩ-impedance of the measurement device and the measured value would be much too small. Therefore, care must be taken to disconnect all intended earth connections such as potential equalization conductors or additional protective earth connectors of medical systems. In addition also unintended connections to earth must be considered such as via data links, mounting screws of permanently installed devices (e.g. ceiling-mounted X-ray C-bows, operating tables, dental chairs), cooling water supply (e.g. power lasers) or accidental contacts such as to supports or even just via a protruding screw of a device stand which contacts the earthed table. Unintentional earthing needs particular attention at
  - direct measurement of enclosure leakage currents and patient leakage currents if earthed voltages are applied;
  - alternative measurements of enclosure leakage currents and patient leakage currents if earth-related measurement voltages are applied instead of using an insulating transformer.
- alternative leakage current measurements rely on contacting the *entire mains* part. However, since the input poles are short-circuited, electronic switches or relays cannot be activated anymore, and contact to circuit parts behind them remains interrupted and, hence, their insulation excluded from testing. Therefore, leakage currents remain too low and insulation failures might remain undetected.
- for differential measurements unintentional earth contacts are irrelevant. However,
  - sensitivity and accuracy of the current clamp could critically degrade measurements compared to other methods.
  - measurement errors could also result from lack of frequency assessment of current harmonics. This is particularly important for devices with electronic

power control where phase clipping can cause a considerable amount of harmonics and frequency assessment becomes relevant. However, only in this case, measurement would overestimate current values and therefore stay on the safe side.

4. Leakage current limits

Enclosure leakage current

Enclosure leakage currents are usually higher than normal condition touch currents as measured according to EN 60601-1. The reason is that measurement of enclosure leakage currents is performed under single fault conditions (e. g. interrupted protective earth conductor). In addition, alternative leakage currents are higher because voltage is now applied to both poles simultaneously while it usually contacts only one of the two poles. For this reason the value of the alternative enclosure leakage current should equal the sum of the two values measured directly or with the differential method with the plug in either position. Therefore, as a principle, results of the alternative measurement method are larger. Consequently, enclosure leakage current limits are defined twice as high as for the other alternatives. If circuits were designed symmetrically, the limit would be exactly twice the enclosure leakage current limit for the other measurement alternatives as defined in EN 60601-1 /27/ (Table 10-10).

Table 10-10: Limits for enclosure leakage currents

Device leakage current	Typ B, BF, CF µA
Safety class I (protective earthing)	
Alternative measurement	1000
Direct measurement	500
Differential measurement	
Safety class II (protective insulation)	
Alternative measurement	500
Direct measurement	100
Differential measurement	

Patient leakage current

The alternative measurement of the patient leakage current equals the single fault condition "mains voltage at applied part." For this case, the standard EN 60601-1 defines a limit value of 5,000 µA (type BF) and 50 µA (type CF), respectively (Table 10-11). Therefore, limits for alternatively and directly measured patient leakage currents have been set accordingly.

Table 10-11: Limits for patient leakage currents

Patient leakage current	Applied part		
	Typ B µA	Typ BF µA	Typ CF µA
alternative measurement	[a]	5000	50
direct measurement	[a]		

[a] Attention! Measurement of applied parts type B causes short-circuiting!

10.10.1.3 Insulation impedance

The measurement of the (high) insulation impedance is performed with a measurement voltage of 500 V. At such a voltage damage to the mains filter or to applied parts can no longer be excluded. Therefore, this measurement should be performed only after it was assured that the device would withstand this procedure. It is not intended as a routine measurement for acceptance of devices (EN 62353). However, it is helpful in the following circumstances

- if there are doubts on sufficient insulation such as after spilling liquids, or if the residual current circuit breaker has been activated several times;
- if it is necessary to monitor changes of the insulation capability with time, such as caused by abrasion of carbon brushes of a DC motor (e. g. at centrifuges);
- if it is suspected that insulation might be sensitive to moisture or to temperature changes.

To measure the insulation impedance the device is separated from the mains. The following kinds of insulation are measured:

- between mains part (Figure 10-18) and
  - protectively earthed accessible metallic parts of safety class I devices (basic insulation);
  - floating accessible metallic parts of devices of safety class I or safety class II (double insulation);
  - all short-circuited patient terminals, irrespective of their type (double insulation);
- between short-circuited (earth-free) applied parts (Figure 10-19) and
  - protectively earthed enclosure (basic insulation);
  - floating metallic enclosure (double insulation).

10.10.2 Function test

Function tests are an important part of recurrent testing. They have several objectives:

1. to reassure that after the test the device is still in the correct status. Since breakdown of components occurs randomly, it could be that it occurs even shortly after testing.

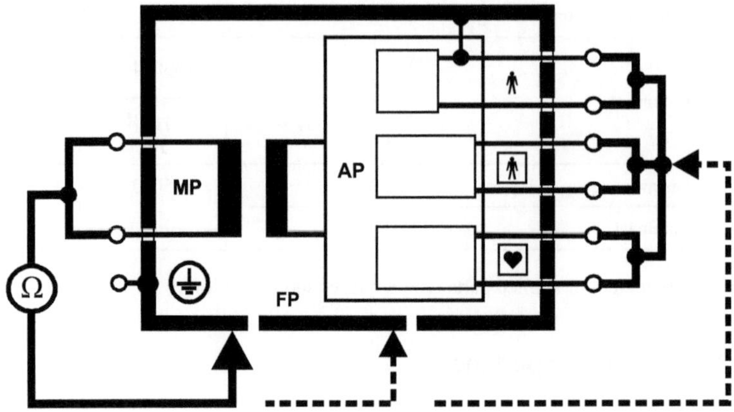

Figure 10-18: Measurement of the insulation impedance between mains part and other parts (the dotted lines indicate multiple measurement positions). MP ... mains part, AP ... applied part(s), FP ... floating metallic part

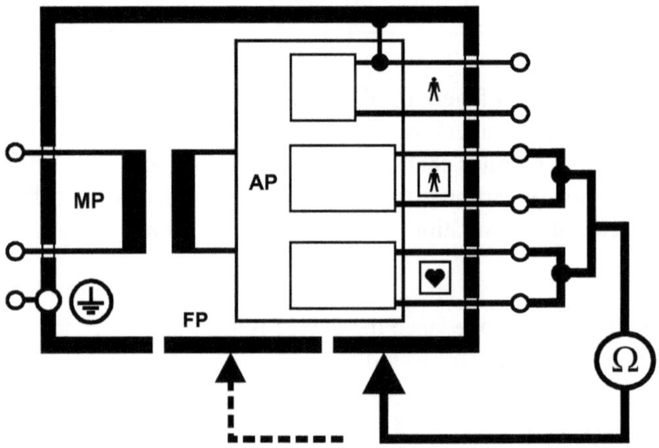

Figure 10-19: Measurement of the insulation impedance between applied part(s) and other parts (the dotted line indicates another measurement position). AP... applied part(s), FP... floating metallic part

Therefore, testers should be sure that this is not due to their work, i.e. because of rough testing of connections or because of measurement of the insulation impedance (with 500 V).

2. to check the functioning of safety-relevant components such as indicator lamps or alarms. This should reassure that technical function rather than accuracy, for example of alarm limits, is still given. The testing of alarm limits should of course be done by the user and in short intervals.

Remark: If after testing the initial settings, for example of alarm limits, have been forgotten or cannot be restored, the user should be informed of this by appropriate means such as by use of a self-adhesive notice. The reason is that in real life physicians don't really check whether settings are still correct prior to using devices.

3. if the function of a device is safety-relevant (e. g. delivery of defibrillation energy, substances or radiation) it has to be checked following the instructions from the manufacturer. It could be necessary to ask for assistance from a person trained in using it. Measurements should be performed to verify whether the output accuracy is still within the given limits. This might require special test equipment such as an infusion pump tester, RF-surgery device tester, laser meter, defibrillator tester or a patient simulator to check patient monitoring devices.

Safety-relevant testing of functions is an indispensable part of recurrent testing. However, it is important to emphasize that it is restricted to safety-relevant aspects only and does not comprise ***all*** functional parameters. General function control (e. g. calibration of ECG recorders, checking the counting rate in nuclear medicine or the operational readiness of a defibrillator) should be done by the user in much shorter intervals than the 2 or 3 years that are foreseen for recurrent testing. It would be a fatal misunderstanding if users believed recurrent testing disburdens them from their own responsibility for functional testing and quality control.

! Recurrent safety testing does not disburden users from their own responsibility of regularly checking a device's function and quality

11 Abbreviations

ALARA	As low as reasonably achievable
ALARP	As low as reasonably practicable
ANSI	American National Standards Institute
BZD	Directive of Biocidal Products
C	Capacitor
CD	Cosmetics Directive
CEN	European Committee for Standardization
CENELEC	European Committee for Electrotechnical Standardization
CNS	Central nervous system
EC	European Community
EEG	Electric encephalogram
EKG	Electric cardiogram
EMG	Electric myogram
EN	European standard
EMA	European Medicines Agency
EES	European Economic Space
FDA	Food and Drug Administration, USA
FI	Residual current circuit breaker
GMDN	Global Medical Device Nomenclature
HD	Harmonization document
I	Electric current
IEC	International Electrotechnical Commission
ISO	International Organization for Standardization
MD	Directive on Machinery
MDD	Medical devices directive
MSELV	Medical safety extra-low voltage
NC	Normal condition
NIST	Non-interchangeable safety thread
ÖVE	Austrian Electrotechnical Association
ON	Austrian Standards Institute
PA	Potential equalization
PBB	Polybrominated Biphenyl
PBDE	Polybrominated Diphenylether
PE	Protective earth
PPE	Directive on Personal Protective Equipment
Q	Electric charge
QM	Quality management
R	Electric resistance
RoHSD	Directive of Restriction of certain Hazardous Substances
RSD	Cleaning, sterilization, disinfection

SFC	Single fault condition
SELV	Safety extra-low voltage
SIP	Signal input part
SOP	Signal output part
U	Electric voltage
UMDNS	Universal Medical Device Nomenclature System
VDE	German Association for Electrical, Electronic & Information Technologies
WEEE	Waste Electrical and Electronic Equipment Directive
Z	Electric impedance

12 Homepages

Information on legal requirements

http://www.europa.eu.int	European Commission
http://eur-lex.europa.eu	Information about European Laws
http://www.bmgfj.gv.at	Austrian Ministry of Health
http://www.bmg.bund.de	German Ministry of Health
http://www.bag.admin.ch	Swiss Federal Office of Public Health
www.fda.gov	US Food and Drug Administration

Information on European market approval

http://ec.europa.eu/enterprise/newapproach/nando	European Commission, Directorate for Enterprises and Industry
http://www.eotc.be	European Organization of Conformity Assessment
http://www.eucomed.be	European Cooperation of Medical Devices Manufacturers
http://www.edma.be	European Diagnostics Manufacturer Association
http://www.pmg.tugraz.at	European Notified Body of Medical Devices (PMG)

Information on standards

Electrotechnical

http://www.cenelec.be	European Committee for Electrotechnical Standardization
http://www.iec.ch	International Electrotechnical Commission
http://www.ove.at	Austrian Electrotechnical Association
http://www.vde.de	German Association for Electrical, Electronic & Information Technologies
http://www.electroswiss.ch	Swiss Association for Electrical Engineering, Power and Information Technologies

General

http://www.cen.be	European Committee for Standardization
http://www.iso.ch	International Organization for Standardization
http://www.on-norm.at	Austrian Standards Institute
http://www.din.de	German Standards Institute
http://www.snv.ch	Swiss Association for Standardization

13 Literature

/1/ Biegelmeier, G. (1986): Effects of electric current on human beings and livestock (German). VDE-Verlag Berlin

/2/ Brinkmann, K., Schaefer, H.: (1982): The electric accident (German). Springer-Verlag Berlin

/3/ BGR 132 (2003): Prevention of fire hazard caused by electrostatic discharges (German). BG-Rule, Association of Commercial and Industrial Workers, Heidelberg

/4/ CENELEC HD 395-1 S2 (1988): Medical electrical equipment, part 1: General requirements for safety

/5/ CENELEC HD 395-1 S1 (1979): Safety of medical electrical equipment. Part 1. General requirements for safety

/6/ DIN 100-107:2002: High-power installation in hospitals and medically used locations outside hospitals (German)

/7/ Directive 2007/47/EC (2007) Amending directives on active implantable medical devices and medical devices

/8/ Directive 2006/42/EC (2006) on Machinery (MD)

/9/ Directive 2002/96/EC (2002) on Waste Electrical and Electronic Equipment (WEEE-D)

/10/ Directive 2002/95/EG (2002) on Restriction of Certain Hazardous Substances (RoHE)

/11/ Directive 2001/83/EC relating to medicinal products for human use

/12/ Directive 1999/44/EC (1999) on certain aspects of the sale of consumer goods and associated guarantees

/13/ Directive 98/79/EC (1998) on in-vitro diagnostic devices (IVD)

/14/ Directive 93/42EC (1993) on medical devices (MDD)

/15/ Directive 90/385/EC (1990) on implantable active medical devices (AIMD)

/16/ Directive 89/686/EC (1989) on personal protective devices (PPD)

/17/ Directive 85/374EC (1984) on liability for defective products

/18/ Directive 76/768/EC (1976) on cosmetic products (CD)

/19/ Ecker, W., Füszl, S., Renhardt, M., Semp R. (2004): Medical devices law (German). Juridica Vienna

/20/ Eikmann, T., Christiansen, B., Exner, M., Herr, C., Kramer, A. (2006): Hygiene in hospital and practice (German). Ecomed Verlag Landsberg/Lech

/21/ EN ISO 14971 (2007): Application of risk management to medical devices

/22/ EN ISO 15001 (2009): Anaesthetic and respiratory equipment – Compatibility with oxygen

/23/ EN ISO/IEC 17020 (2004): General criteria for the operation of various types of bodies performing inspections

/24/ EN ISO/IEC 17025 (2005): General requirements for the competence of testing and calibration laboratories

/25/ EN IEC 60335-1 (2004): Household and similar electrical appliances – Safety. Part 1: General requirements

/26/ EN IEC 60479-1 (2007): Effects of current on human beings and livestock – Part 1: General aspects

/27/ EN 60601-1, Ed.3 (2006): Medical electrical equipment – Part 1: General requirements for basic safety and essential performance

/28/ EN IEC 60601-1-2 (2006): Medical electrical equipment – Part 1-2: General requirements for basic safety and essential performance. Collateral standard: Electromagnetic compatibility – Requirements and tests

/29/ EN IEC 60601-1-6 (2008): Medical electrical equipment – Part 1-6: General requirements for safety – Collateral standard: Usability

/30/ EN IEC 60601-1-8 (2004): Medical electrical equipment – Part 1-8: General requirements for safety – Collateral standard: Alarm Systems – General requirements, tests and guidance for alarm systems in medical electrical equipment and medical electrical systems

/31/ EN IEC 60601-1-9 (2006): Medical electrical equipment – Part 1-9: General requirements for basic safety and essential performance – Collateral standard: Requirements for environment-conscious design

/32/ prEN IEC 60601-1-11 (2009): Medical electrical equipment – Part 1-11: General requirements for basic safety and essential performance – Collateral standard: Requirements for medical devices and medical electrical systems for home use

/33/ EN IEC 61000-4-8 (2001): Electromagnetic compatibility (EMC) – Part 4-8: Testing and Measurement Techniques – Power Frequency Magnetic Field Immunity Test

/34/ EN IEC 62304 (2006): Medical device software – Software life-cycle process

/35/ EN IEC 62353 (2008): Medical electrical devices: Recurrent test and test after repair of medical electrical equipment

/36/ FDA (2004): Federal Food, Drug and Cosmetic Act. USA, http://www.fda.gov/opacom/laws/fdcact/fdctoc.htm

/37/ Gärtner A. (2008): Safety of medical products. Part 2: Electrical safety in medical technology (German). TÜV Rheinland Köln

/38/ Haase, H. (1972): Dangerous static electricity (German). Verlag Chemie Weinheim

/39/ Hofheinz, W. (2005): Electric safety in medically used locations: Power supply conforming to standards, and professional testing of medical electrical devices (German). VDE-Verlag Berlin

/40/ Hutten, H. (Ed.) (1990): Biomedical Engineering, (German) Part 1-4. Springer Verlag Berlin

/41/ IEC 601-1, Ed.2 (1988): Medical electrical equipment. Part 1: General Requirements

/42/ IEC 601-1, Ed.1 (1977): Safety of medical electrical equipment. Part 1: General Requirements

/43/ IEC 60601-1, Ed.3 (2005): Medical electrical equipment – Part 1: General requirements for basic safety and essential performance

/44/ ISO 32 (1977): Gas cylinders for medical use – Marking for identification of content

/45/ Kramme, R. (Ed.) (2001): Medical Technology (German). Springer-Verlag Berlin

/46/ Leitgeb, N., Omerspahic, A., Niedermayr, F. (2009): Exposure of non-target tissue in medical diathermy. Bioelectromagnetics, 30:Adv.Publ. Aug.29,2009

/47/ Leitgeb, N., Cech, R., Schröttner, J., Lehofer, P., Schmidpeter, U., Rampetsreiter, M. (2008): Magnetic emission ranking of electric appliances. A comprehensive market survey. Radiat. Prot. Dosim., 129:439-445

/48/ Leitgeb, N., Schröttner, J., Cech. R. (2007): Perception of ELF electromagnetic fields: Excitation thresholds and inter-individual variability, Health Physics, 92(6):591-595

/49/ Leitgeb, N., Schröttner J., Cech R. (2005): Electric current perception of the general population including children and elderly. J. Med. Eng. Technol. 29, 215-218

/50/ Leitgeb, N. (2000) Do electromagnetic fields cause illness? (German). Springer Verlag Berlin

/51/ Leitgeb, N. (1996): Safety in Electromedical Technology. Interpharm Press, Buffalo Grove, IL

/52/ Leitgeb, N. (1990): Rays, Waves, Fields (German). Thieme Verlag Stuttgart

/53/ Medical Devices Law (1996). Austrian Law Gazette. BGBl 212

/54/ Ordinance on Medical Devices Operators (2007). Austrian Law Gazette. BGBl 22

/55/ ÖVE/ÖNORM E 8007 (2007): High-power installation in hospitals and medically used locations outside hospitals (German)

/56/ Perrow, C. (1989). Common catastrophes (German). Campus Verlag, New York

/57/ Product Liability Law (1988). Austrian Law Gazette. BGBl 99

/58/ Roy O. Z., Scott, J. R., Park G. C. (1976): 60Hz ventricular fibrillation and pump failure thresholds versus electrode area. IEEE Trans BME `23, 45-48

/59/ Starmer C. F., Watson R. E. (1973): Current density and electrically induced ventricular fibrillation. Med. Instr. 7, 3-6

/60/ VDE 0100-710 (2002): Installation of low voltage facilities – Requirements for industrial sites and locations and installations of a special kind. Part 710: Medically used locations (German)

14 Figures

15 Tables

15 Subject Index

MIX
Papier aus verantwortungsvollen Quellen
Paper from responsible sources
FSC® C105338

If you have any concerns about our products, you can contact us on
ProductSafety@springernature.com

In case Publisher is established outside the EU, the EU authorized representative is:
Springer Nature Customer Service Center GmbH
Europaplatz 3, 69115 Heidelberg, Germany

Printed by Libri Plureos GmbH
in Hamburg, Germany